D0875798

Electric Vehicle Engineering

Per Enge

Nick Enge

Stephen Zoepf

New York   Chicago   San Francisco
Athens   London   Madrid
Mexico City   Milan   New Delhi
Singapore   Sydney   Toronto

Library of Congress Control Number: 2020947347

Electric Vehicle Engineering

1 2 3 4 5 6 7 8 9   CCD   25 24 23 22 21 20

ISBN 978-1-260-46407-8
MHID 1-260-46407-5

This book is printed on acid-free paper.

Sponsoring Editor	**Copy Editor**
Lara Zoble	Nandita Singha, MPS Limited
Editorial Supervisor	**Proofreader**
Stephen M. Smith	Syed Zakaullah, MPS Limited
Production Supervisor	**Indexer**
Pamela A. Pelton	Edwin Durbin
Acquisitions Coordinator	**Art Director, Cover**
Elizabeth M. Houde	Jeff Weeks
Project Manager	**Composition**
Rishabh Gupta, MPS Limited	MPS Limited

For Per,
who provided the essential electromotive force
that propelled this project and so many others

About the Authors

Per Enge, Ph.D., M.S., B.S., was Director of the GPS Lab at Stanford University. He was a Member of the National Academy of Engineers and GPS Hall of Fame, a Fellow of ION and IEEE, and author of *Global Positioning System: Signals, Measurements, and Performance.*

Nick Enge, M.S., B.S., developed and co-taught Stanford University's first modern undergraduate course on electric vehicles. He is currently a lecturer at the University of Texas at Austin and is co-author of *The Science of Speaking.*

Stephen Zoepf, Ph.D., M.S., B.S., is the Chief of Policy Development for Ellis & Associates, where he helps guide the development of open-source software products for cities to manage modern transportation systems. He teaches at Stanford University and has two decades of experience in transportation and mobility.

Contents

Preface

The seed of this book was planted in 2009, when Don Cox, a professor of electrical engineering at Stanford University, offered Per Enge, a professor of aeronautics and astronautics, and his son, Nick Enge, a recent graduate in atmosphere/energy engineering, test drives in his Tesla Roadster. ("Don't worry about the car," Don said, "But any speeding tickets are yours.") As soon as Per and Nick felt the power of the Roadster's 185-kW induction motor accelerating the car from 0 to 60 mph in under 4 sec, they were hooked.

The following school year, Per and Nick collaborated with Don to develop an introductory seminar for Stanford sophomores to explore the technology that enabled such a marvel of engineering. Don provided the electrical perspective, as well as engaging in some "unbiased" electric vehicle evangelism. Per grounded everything in physics, and Nick contributed his energy and environment expertise. When Don retired, Per and Nick soldiered on, continuing to teach the course in its original form until 2017, when Stephen Zoepf joined the teaching team and helped them transform the course from an introductory seminar to a regular sophomore-level course offering in Mechanical Engineering, while adding a new policy perspective.

This book was written as a textbook for that course. Although its publication was delayed by Per's passing in 2018, Nick and Stephen are pleased to be able to bring Per's vision across the finish line with the help of McGraw Hill. We hope that you will find as much benefit in the information it provides as nearly a decade of our Stanford students have.

Before we continue, it's important to note several ways in which this book differs from many of the other electric vehicle textbooks on the market.

First, given that it was written for college sophomores, the only prerequisite for understanding its entire contents is a basic knowledge of physics: an introductory college physics series, or Physics AP in high school (if it included E&M). While all three authors have advanced degrees in science and engineering, the entire book has been reviewed by a variety of test readers from different backgrounds, including Melissa Enge, Alfred Zoepf, Lee Carvell, and the students of ME182 in Fall 2018, who have all confirmed that this is indeed the only prerequisite for understanding. For this reason, we believe that this book will be of interest not only to undergraduate engineering students but also to owners of electric vehicles who want to better understand how they function.

Second, given that it was written by three engineers, our perspective is grounded in real-world data, and we take a truly hands-on approach to the subject. This is most clearly evident in Chap. 3, where we use Nick's Tesla Model 3 as a testbed to validate our theoretical model for the energy it consumes, and in the final project for our course, presented in App. A, where students design their own model electric car with a motor they have built from the ground up with their own hands.

Finally, it is one of very few electric vehicle books that lives fully in the new age of EVs, where lithium-ion batteries enable ranges upward of 200 miles. You won't find examples here based on the EV1 (which has been off the road for nearly two decades now), or other outdated technologies like lead-acid batteries. Instead, most examples in this book are based on the Tesla Model 3, which we believe to be a good model for where the field is going, not where the field has been.

It is our hope that this book can serve as a basis for other courses like the one it was designed for at Stanford and that it can contribute to the education of a new generation of electric vehicle engineers, of whom we will need many more in the coming decades. Here is a brief overview of what we will cover:

In Chap. 1, we'll discuss the promise of electrification, a development which the National Academy of Engineers declared to be the greatest engineering achievement of the twentieth century, and show how it will apply to transportation.

In Chap. 2, we'll explore the history of electric transportation. While electric vehicles are often considered a modern technology, they have actually been around for more than 150 years. In this chapter, we'll take a look at some of the major milestones in their development.

In Chap. 3, we'll dive into the science of vehicle dynamics. Building up from first principles in mechanics, we will develop an answer to the essential question: how much energy does it take to get around? Given that wheeled vehicles account for 84 percent of transportation energy use in the United States, we will focus on the energy required for this type of vehicle. In this chapter, we will build and validate a model that tells us how much energy a given vehicle will require to travel along a particular route, and use it to predict energy consumption, efficiency, and range.

In Chap. 4, we'll see how electric motors allow us to convert electricity into vehicle motion. Building up from first principles in electromagnetism, we will explore the workings of a variety of different motor types including brushed motors, brushless motors, reluctance motors, and induction motors. In addition, we'll develop an expression for motor efficiency, and a simple formula that can be used to predict the maximum acceleration of an electric car using nothing more than its weight and the power of its motor.

In Chap. 5, we'll see how batteries work to store electricity for later use by the motor. Given that most electric vehicles today run on lithium-ion batteries, we'll focus on this particular chemistry. In addition, we'll explore electric vehicle charging in order to understand how electricity makes it from the power grid into the battery.

In Chap. 6, we'll see how controllers are used to convert the electricity stored in the battery into electricity that can be used to run the motor. Building up from first principles in circuit design, we will explore the design of step-down DC controllers, step-up/down DC controllers, and AC controllers. Along the way, we will learn how pulse width modulation is used to generate higher or lower DC voltages and to turn constant DC voltage into AC voltage with variable frequency.

In Chap. 7, we'll take a look at a wide variety of fuel sources that could be used to power transportation and the technologies that could be used to convert those fuels into miles traveled. In doing so, we'll compare electric vehicles to other kinds of vehicles with an eye toward determining which technology is most efficient. Spoiler alert: it's almost always the electric vehicle. In addition, we will compare the emissions associated with driving an electric vehicle to the emissions of traditional gasoline-powered vehicles. Once again, we will see that in most cases, the electric vehicle is the winner.

In Chap. 8, we'll take a look at some of the incentives and barriers influencing electric vehicle adoption. Specifically, we'll discuss a variety of different policies that governments around the world have successfully used to increase the number of EVs on the road, and talk about two of the biggest challenges that the EV market has faced, namely range anxiety and cost.

Finally, in Chap. 9, we'll look toward the future and discuss some of the more recent developments in electric vehicle technology and how they might further revolutionize the transportation industry going forward.

Along the way, we'll present a variety of homework problems and other assignments to further cement your understanding of the material. At the end of the book, in App. A, we will present a long-form project that involves building a model electric car from scratch. After reading this book and completing this project, you will truly be able to call yourself an electric vehicle engineer.

Nick Enge
Stephen Zoepf

CHAPTER 1

The Promise of Electrification

Whether by car, train, boat, or plane, we all need to get around. And we certainly do get around: on U.S. highways alone, we traveled 3.3 trillion miles in 2019—to Pluto and back over 300 times.[1] Of course, to do so, we use a lot of energy: to travel all of those miles, we burned almost half a trillion gallons of fuel.[2] If the Empire State Building was our national fuel tank, it would be emptied nearly five times per day, at a rate of 20 stories of fuel per hour.

The vast majority of this fuel is, in fact, fuel: 92 percent of the energy used for transportation in the United States comes directly from petroleum.[3] And although we often take this fact for granted, when we compare transportation to the other sectors of the economy (industrial, residential, and commercial), we see that this pattern is actually quite odd.[a]

As Fig. 1.1 illustrates, transportation is a glaring outlier: it is far and away the most homogeneous of all sectors in terms of the energy resources it relies on. In the industrial, residential, and commercial sectors, a wider variety of fuel is used, and a large and increasing proportion of the energy required is delivered in the form of electricity.

At the turn of the twentieth century, less than 2 percent of the energy in the United States was provided by electricity: the other 98 percent was provided by the direct burning of fuel.[5] Today, 37 percent of our energy is provided by electricity.[b] In 2000, the National Academy of Engineering identified this shift toward electrification as the "greatest engineering achievement of the 20th century."[7]

Of course, this shift is far from over, as electrification continues to increase around the world. Even in the United States, where we have been thoroughly electrified for a long time, the share of electricity in

---

[a]Transportation accounts for 28 percent of total energy use in the United States, compared to 32 percent for industrial, 21 percent for residential, and 18 percent for commercial.[4]

[b]This figure includes the transportation sector. Excluding transportation, electricity accounts for 51 percent of our energy use (in the industrial, residential, and commercial sectors).[6]

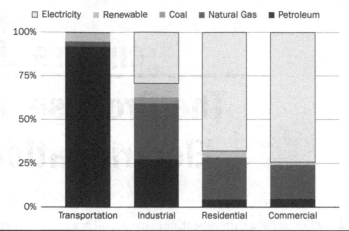

FIGURE 1.1 Energy for transportation is unusually homogeneous.

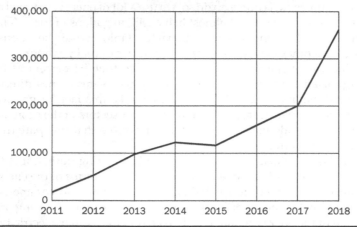

FIGURE 1.2 Yearly electric vehicle sales in the United States.

the energy mix of the commercial, residential, and industrial sectors continues to increase.[8] This begs the question: why hasn't transportation gone the same way?

The answer is that it's *going* the same way—it's simply still in the early stages. Over the past decade, we've seen an explosion in electric vehicle sales, both in the United States and around the world. From less than 20,000 electric cars sold in the United States in 2011 to more than 360,000 in 2018, the market for electric vehicles has grown by almost a factor of twenty in less than a decade (Fig. 1.2). Around the world, the numbers are even more impressive: more than 2 million electric cars were sold in 2018.[9]

In both cases, this represents about 2 percent of total car sales. At first, this may not seem like a lot, but remember, electricity provided a similarly small percentage of total energy in 1900, only to grow like wildfire in the next few decades. Is it possible that the same thing will happen for transportation? And if it does, will that be a good thing?

In answering these questions, it's useful to consider why electrification has occurred in the residential, commercial, and industrial sectors over the past century. As noted before, at the turn of the twentieth century, energy was provided almost exclusively by the direct burning of fuel, as it had been for countless centuries before.

In the late nineteenth century, Thomas Edison developed the first commercially viable electric lightbulb, and power stations and power grids sprouted up to support it. As a method of lighting, electricity represented a great improvement over the gaslight that preceded it. Electric lights were brighter, more reliable, cleaner, and safer than gaslights. In the beginning, electricity spread because it improved existing services.

As the availability of electricity increased, so did its number of potential uses. The first few decades of the twentieth century saw the invention of the electric vacuum cleaner, refrigerator, dishwasher, air conditioner, and washer and dryer for clothes. Thus, in addition to improving upon existing technologies, electricity enabled completely new ones. Today, we rely on electricity for almost everything: from the television to the Internet to the mobile phone, none of these would be possible without electricity.

These are the most obvious benefits of electricity, the ones we as consumers see every day. But it turns out that there are many more if we pull back the curtain and look behind the scenes. For example, one of electricity's most revolutionary qualities is that it decouples the source of energy from its end use.

On the demand side, we've seen that electricity allows us to do a wide variety of things: whether our goal is to light or cool or clean a room, we can do it all from a single socket.[c] A single source can power hundreds of different things.

On the supply side, it's the same story: hundreds of different power generation methods can be used to turn a wide variety of fuels into electricity. Whether it's fossil fuels like coal, oil, and natural gas, nuclear fission, or renewable forms of energy such as solar, wind, and geothermal, all of these energy sources can be used to move electrons, which can, in turn, power any electrical device. In this way, electricity acts like a universal translator. Almost any energy source can be used to power almost any device, as long as you turn it into electricity first.

There are many different benefits to this setup. First, it means that you don't need to transport the raw fuel all the way to its point of use. You simply need to get it to the power generation facility, then

---

[c] As a historical aside, given that the first use of electricity was lighting, early appliances all plugged into light sockets. And given that light sockets were often mounted in the ceiling, other early appliances were often plugged in there as well—sometimes directly connected to the chandelier—which resulted in a spiderweb of electric lines snaking throughout the room.

transport electricity the rest of the way. Relatedly, given that the fuel use happens in one place, rather than being widely distributed, any negative effects—like pollution—can be more carefully contained.

In addition, electrification can make our energy system more resilient. While many people talk about achieving "energy independence," which means being able to support our energy needs without relying on other countries, former Intel CEO Andy Grove has proposed that a better goal might be "energy resilience."

In a globalized energy market, Grove says, it's unrealistic to believe that we will ever be energy independent in the purest sense of that phrase, and even if we could be, this might not be the best way forward. Instead, we should build a resilient energy system, one that is resistant to the kinds of shocks that energy independence aims to protect us from—like the oil crises of the 1970s—without requiring that we actually be independent.

As Grove notes, electricity is resilient for two reasons. As we saw before, it's multi-sourced: even a power plant that was originally designed to be powered by a specific fuel type can often be retrofitted to run on something else, if necessary. But it also has a quality that Grove calls "stickiness"—electricity tends to stay close to where it is produced, instead of being shipped around the world. As Grove writes, "Because electricity is the stickiest form of energy, and because it is multi-sourced, it will give us the greatest degree of energy resilience. Our nation will be best served if we dedicate ourselves to increasing the amount of our energy that we use in the form of electricity."[10]

These are just a few of the reasons that electrification deserves the title of the "greatest engineering achievement of the 20th century." But the question is: can these same benefits translate to electric transportation? The answer, as we'll see, is yes, they can, and as the electrification of transportation increases, they will. Just as electrification has revolutionized the residential, commercial, and industrial sectors, electric transportation has the potential to improve upon existing technologies, enable completely new ones, and decouple the source of energy from its end use.

Electric transportation has the potential to radically improve upon our current experience. For example, for those who have the need for speed, the Tesla Model S is the fastest-accelerating sedan in history, while being relatively cheap for that level of performance, and astronomically more efficient. On that note, electric cars of all different types have the potential to reach new records in efficiency, using less energy than ever before. As we will see in Chap. 7, electric vehicles are the most efficient way to turn almost any fuel into miles travelled, even without any further improvements in technology.

In addition, electric transportation has the potential to open up completely new possibilities. Imagine that you could simply drive into your garage at night and your car would always be ready to go the next morning, without ever having to go to the gas station. With new charging technologies like Plugless Power's inductive charging mat (currently available), you never have to worry about refueling your car: all you have to do is get in and drive. And it's possible that in the future, you could even charge your car on the highway, simply by driving down a special inductive charging lane.

Finally, by decoupling the source of energy from its end use, the electrification of transportation has the potential to break transportation's dependence on oil and allow us to use other fuels for this purpose, just as they are already used in the other three industries. The end result will be a transportation system that is much more resilient.

Notes

1. U.S. Federal Highway Administration. "Moving 12-Month Total Vehicle Miles Traveled," retrieved from FRED, Federal Reserve Bank of St. Louis. https://fred.stlouisfed.org/series/M12MT VUSM227NFWA

2. Davis, Stacy C., and Robert G. Boundy. "Table 1.14. Highway Transportation Petroleum Consumption by Mode, 1970–2017." *Transportation Energy Data Book (Edition 38)*. Oak Ridge, TN: Oak Ridge National Laboratory, 2020. https://tedb.ornl.gov/data/

3. Davis, Stacy C., and Robert G. Boundy. "Table 2.2. Distribution of Energy Consumption by Source and Sector, 1973 and 2018." *Transportation Energy Data Book (Edition 38)*. Oak Ridge, TN: Oak Ridge National Laboratory, 2020. https://tedb.ornl.gov/data/

4. Davis, Stacy C., and Robert G. Boundy. "Table 2.1. U.S. Consumption of Total Energy by End-Use Sector, 1950–2018." *Transportation Energy Data Book (Edition 38)*. Oak Ridge, TN: Oak Ridge National Laboratory, 2020. https://tedb.ornl.gov/data/

5. Institute for Energy Research. "History of Electricity." http://instituteforenergyresearch.org/history-electricity/

6. Calculated by combining the data in Tables 2.1 and 2.2 from: Davis, Stacy C., and Robert G. *Transportation Energy Data Book (Edition 38)*. Oak Ridge, TN: Oak Ridge National Laboratory, 2020. https://tedb.ornl.gov/data/

7. National Academy of Engineering. "Greatest Engineering Achievements of the 20th Century." http://www.greatachievements.org

8. Davis, Stacy C., and Robert G. Boundy. "Table 2.2. Distribution of Energy Consumption by Source and Sector, 1973 and 2018." *Transportation Energy Data Book (Edition 38)*. Oak Ridge, TN: Oak Ridge National Laboratory, 2020. https://tedb.ornl.gov/data/

9. Inside EVs. "Quarterly Plug-In EV Sales Scorecard." http://inside evs.com/monthly-plug-in-sales-scorecard/

10. Grove, Andy. "Our Electric Future." *The American*, July 10, 2008. http://www.aei.org/publication/our-electric-future/

Image Credits

Figures 1.1 and 1.2: Nick Enge

CHAPTER **2**

History of
Electric Vehicles

2.1 Introduction

While electric vehicles are often considered a modern technology, they have actually been around for quite a long time: more than 150 years, in fact. Here are some of the most interesting events in the history of electric vehicles.

2.2 Early History

c. 3,500 BCE The first evidence and illustrations of wheeled vehicles appear. (Fig. 2.1)

c. 1680 Flemish missionary Ferdinand Verbiest designs the first self-moving vehicle, or "automobile" (from the Greek *autos* "self," and the French *mobile* "moving"), powered by a steam turbine, as a toy for the Chinese Emperor. (Fig. 2.2)

FIGURE 2.1 From the Standard of Ur, c. 2600 BCE.

FIGURE 2.2 Verbiest's miniature steam-powered automobile.

FIGURE 2.3 Cugnot's full-sized steam-powered automobile.

1749 Benjamin Franklin first uses the term "battery" to describe a group of capacitors linked together to produce a stronger discharge. Prior to this, "battery" simply referred to a group of things working together, as in an artillery battery.

1769 French Army captain Nicholas Cugnot develops a model automobile powered by a reciprocating steam engine. The next year, he builds a full-size version, capable of carrying four people at a speed of 2.25 mph. (Fig. 2.3)

2.3 Nineteenth Century

1800 Stacking metals and brine-soaked cloth in series, Alessandro Volta develops the voltaic pile, the first battery capable of supplying a continuous current to a circuit. (Fig. 2.4)

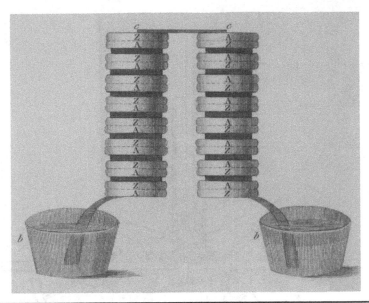

Figure 2.4 Volta's original voltaic pile.

1807 French inventor François Isaac de Rivaz designs a hydrogen-powered internal combustion engine (ICE) and uses it to power a small carriage—the first automobile with an ICE. In 1813, he designs a much larger vehicle, 6 m long, weighing almost a ton. It is loaded with 700 lb of stone and wood, along with four men, and runs for 26 m up a hill. Despite his successes, most people still believe steam power will reign supreme. (Fig. 2.5)

1815 Czech professor Josef Božek builds a steam car powered by oil. (Fig. 2.6)

1820 During a lecture, Danish physicist Hans Christian Ørsted notices that a compass needle is deflected when a nearby current is switched on and off, confirming a relationship between electricity and magnetism. (Fig. 2.7)

1820 André-Marie Ampère invents the solenoid, showing that a uniform magnetic field is produced inside a cylindrical coil of current-carrying wire. (Fig. 2.8)

1821 English scientist Michael Faraday builds the first motor, a homopolar motor in which a current-carrying wire that is extended into a pool of mercury rotates around a magnet. (Fig. 2.9)

1824 English physicist William Sturgeon invents the electromagnet. He demonstrates its power by lifting 9 lb with a small piece of iron wrapped with wire, powered by a battery. (Fig. 2.10)

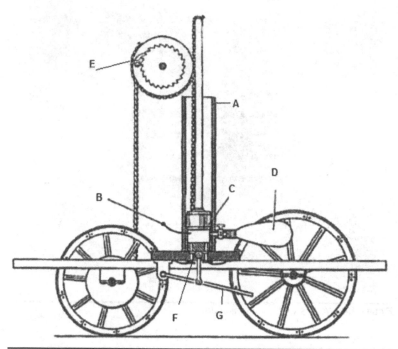

FIGURE 2.5   Rivaz's original internal combustion engine vehicle.

FIGURE 2.6   Replica of Božek's oil-powered steam car.

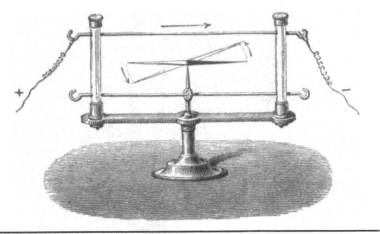

FIGURE 2.7 Ørsted's compass deflection experiment.

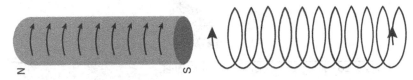

FIGURE 2.8 Ampere's solenoid.

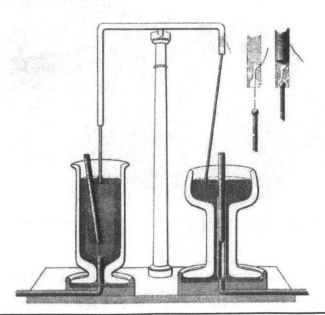

FIGURE 2.9 Faraday's homopolar motor.

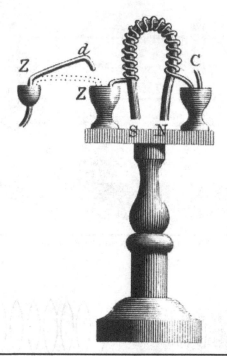

FIGURE 2.10   Sturgeon's original electromagnet.

FIGURE 2.11   Jedlik's original DC motor.

1828   Hungarian priest Ányos Jedlik invents the first direct-current (DC) motor as we know it, complete with stator, rotor, and commutator. He builds a model electric car to demonstrate its potential. (Fig. 2.11)

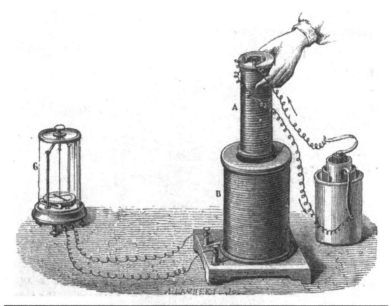

FIGURE 2.12 Faraday's induction experiment.

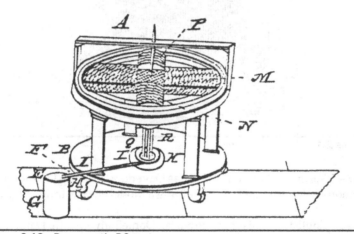

FIGURE 2.13 Davenport's DC motor.

1831 Faraday investigates induction. Current from a battery is used to induce a magnetic field around a coil of wire (A), which, when moved inside another coil of wire, induces a current in the other wire (B), which is detected by a galvanometer (G). (Fig. 2.12)

1833 Russian physicist Heinrich Lenz formulates Lenz's Law, which states that an induced current flows in a direction that opposes the change that produced it.

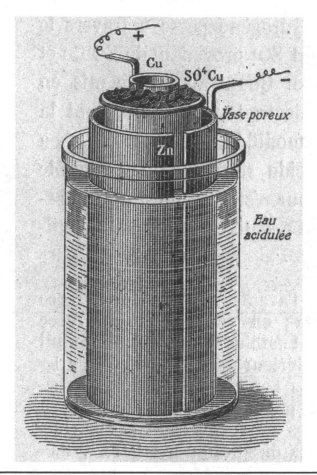

FIGURE 2.14   The Daniell cell.

1834   Vermont blacksmith Thomas Davenport builds a DC motor and demonstrates its potential by building a model electric train. (Fig. 2.13)

1836   English chemist John Frederic Daniell invents the Daniell cell, which uses two metal electrodes and two electrolytes separated by a porous earthenware barrier to produce a longer-lasting and more reliable current than the voltaic cell. (Fig. 2.14)

1838   Prussian engineer Moritz von Jacobi builds a 28-ft electric paddle-wheel boat powered by zinc primary batteries ("primary" in the context of batteries means non-rechargeable). It is capable of transporting a dozen passengers against the current of the Neva River. (Fig. 2.15)

1842   Scottish inventor Robert Davidson builds the first full-size electric locomotive, Galvani, which travels at 4 mph using zinc primary

FIGURE 2.15   Jacobi's electric motor.

batteries. Fearing competition from electric locomotives, steam engineers smash Galvani to pieces in its shed. (Fig. 2.16)

1859   Building on experiments by Wilhelm Sinsteden, French physicist Gaston Planté demonstrates the first practical lead-acid battery. It is the first secondary battery, which means it can be recharged by passing a reverse current through it. (Fig. 2.17)

1861   Scottish mathematician James Clerk Maxwell reduces electromagnetic knowledge to an elegant series of equations.

1867   Austrian inventor Franz Kravogl demonstrates an electric bicycle at the World Exposition in Paris, but it can't drive reliably, and is regarded as a curiosity.

1878   Amédée Bollée produces 50 copies of his steam car, La Mancelle, making it the first automobile put into "mass" production. (Fig. 2.18)

1879   Rotating a copper disk by manually switching the direction of current in an electromagnet, English physicist Walter Baily demonstrates the basic principle of induction motors.

FIGURE 2.16   Advertisement for exhibition of Davidson's Galvani.

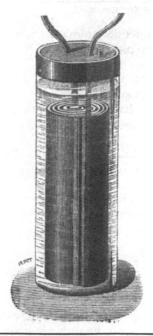

FIGURE 2.17   Planté's 1859 lead-acid cell.

FIGURE 2.18   Amédée Bollée La Mancelle.

FIGURE 2.19   The Trouvé electric tricycle.

1881 French electrical engineer Gustave Trouvé fits an electric motor to a tricycle, making the first manned electric road vehicle. (Fig. 2.19)

FIGURE 2.20 The Lichterfelde electric tramway.

FIGURE 2.21 The Electromote, with overhead power lines.

1881 The first electric tramway is built by Siemens & Halske in Lichterfelde, a suburb of Berlin. Each car is propelled by a 4-kW DC motor powered by the rails. (Fig. 2.20)

1882 The Electromote, also built by Siemens & Halske, is the first electric vehicle to be run like a trolleybus, with electric power supplied by overhead cables. (Fig. 2.21)

FIGURE 2.22   The Benz Patent-Motorwagen.

1885   Italian physicist Galileo Ferraris builds the first induction motor.

1886   Karl Benz builds the Benz Patent-Motorwagen, the first automobile with a gasoline-powered ICE. In 1888, his wife Bertha and their two sons drive 60 mi to visit her mother, the first long-distance road trip in an automobile. Along the way, Bertha invents brake pads. (Fig. 2.22)

1886   Frank Julian Sprague invents regenerative braking. By turning the motor into a generator, an electric vehicle can be slowed down. The energy generated can be used to recharge the battery and increase range.

1887   Nikola Tesla independently invents and patents the induction motor. (Fig. 2.23)

1888   German inventor Andreas Flocken develops the Flocken Elektrowagen, the first electric car to be widely known. (Fig. 2.24)

1891   Russian engineer Mikhail Dolivo-Dobrovolsky invents the so-called "squirrel-cage" induction motor, a popular design still in use today. (Fig. 2.25)

1896   In the first American automobile track race, an electric car wins all five heats.

1897   The Columbia Electric Phaeton Mark III is the first electric car produced in non-trivial numbers. (Fig. 2.26)

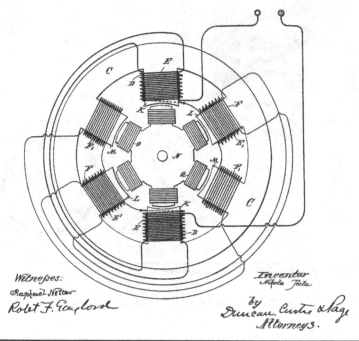

FIGURE 2.23   Tesla's induction motor.

FIGURE 2.24   The Flocken Elektrowagen.

December 18, 1898 Automobiles have become fast enough that French automobile magazine *La France Automobile* holds a contest to establish an official world record for land speed. A French electric car, the Jeantaud Duc, driven by Count Gaston de Chasseloup-Laubat,

FIGURE 2.25 Dolivo-Dobrovolsky's squirrel cage induction motor.

FIGURE 2.26 The Columbia Electric Phaeton Mark III.

sets the first official land speed record of 39.24 mph. Electric cars continue to dominate automobile races throughout 1899 (see Table 2.1). (Fig. 2.27)

1899 Swedish inventor Waldemar Jungner invents the nickel-cadmium battery, with nickel and cadmium electrodes in a potassium

Year/Location	Speed	Car/Driver	Type
12/18/1898: Achères, France	39.24 mph	Jeantaud Duc Gaston de Chasseloup-Laubat	Electric
1/17/1899: Achères, France	41.42 mph	CGA Dogcart Camille Jenatzy	Electric
1/17/1899: Achères, France	43.69 mph	Jeantaud Duc Gaston de Chasseloup-Laubat	Electric
1/27/1899: Achères, France	49.93 mph	CGA Dogcart Camille Jenatzy	Electric
3/4/1899: Achères, France	57.65 mph	Jeantaud Duc Profilée Gaston de Chasseloup-Laubat	Electric
4/29/1899: Achères, France	65.79 mph	La Jamais Contente Camille Jenatzy	Electric
4/13/1902: Nice, France	75.06 mph	Oeuf de Pâques Leon Serpollet	Steam
11/5/1902: Ablis, France	76.08 mph	Mors Z Paris-Vienne William K. Vanderbilt	ICE
All later records were ICE, steam, or turbojet			

TABLE 2.1 Early Land Speed Records

FIGURE 2.27 Count Gaston de Chasseloup-Laubat driving the Jeantaud Duc.

hydroxide solution (the first alkaline electrolyte). While it has better energy density than lead-acid, it is also much more expensive.

April 29, 1899 The first vehicle of any kind to reach 60 mph is an electric car, La Jamais Contente ("The Never Satisfied"), driven by Le Diable Rouge ("The Red Devil"), Camille Jenatzy, so-named for the color of his beard. The car has an aerodynamic, torpedo-shaped body

Figure 2.28 Camille Jenatzy driving La Jamais Contente.

made of aluminum alloy, although much of the aerodynamic benefit is ruined by the driver sticking out the top. Driven by two 25-kW motors, it tops out at 65.8 mph. While this record will hold for the next three years, it is the last general speed record set by an electric car—all later land speed records are set by internal combustion, steam, or turbojet engines. (Fig. 2.28)

1899   Henri Pieper of Germany develops a parallel hybrid car, which uses an electric motor to supplement the power provided by its weak ICE and batteries charged by the engine while coasting and descending hills. The Pieper parallel hybrid starts in electric mode, making it the first ICE vehicle with an electric starter. This is important because until the advent of the self-starter in the early twentieth century, most ICE vehicles rely on hand cranking to start the engine. Hand cranking is a dirty process and presents a non-trivial risk of breaking your arm if the engine backfires, a risk you must face every time you want to start your car. (Fig. 2.29)

1899   Vedovelli, Priestley & Co. develops a series hybrid car, which tows a 0.75-hp ICE coupled to a 1.1-kW generator in order to recharge the batteries on the go and extend range.

2.4   Twentieth Century

1900   At the turn of the century, 40 percent of American automobiles are powered by steam, 38 percent by electricity, and 22 percent by internal combustion engine.

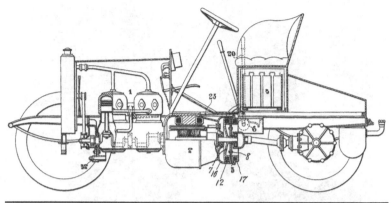

FIGURE 2.29   Illustration of the Pieper's parallel hybrid.

FIGURE 2.30   The Lohner-Porsche Mixte Hybrid.

1900   The Lohner-Porsche Mixte Hybrid, a series hybrid, breaks the Austrian land speed record of 37 mph. (The world record is still Jenatzy's 65.8 mph.) (Fig. 2.30)

1902   The electric Baker Torpedo is the first car designed to be fully aerodynamic and the first car to have seat belts. In an attempt to break the speed record, it exceeds 70 mph before hitting streetcar tracks and veering out of control, plowing through the crowd, killing one and injuring several. (Fig. 2.31)

1903   Wireless Auto No. 1 is the first car (an EV) equipped with radio. It is used for broadcasting stock quotations.

FIGURE 2.31 The Baker Torpedo.

FIGURE 2.32 The Ford Model T.

1908 The Baker electric roadster has an advertised range of 100 mi. Two years later, a Baker electric sets an electric vehicle distance record of 244.5 mi on a single charge.

1908 Ford releases the Model T, the first truly affordable automobile. It runs on a gasoline-powered ICE. (Fig. 2.32)

1911 The first practical self-starter for the ICE is invented by Charles F. Kettering, eliminating the need to hand crank your engine. This is one of the key developments that will allow ICEs to rise in popularity

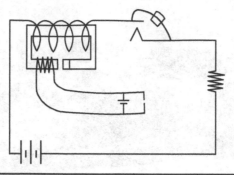

FIGURE 2.33   Kettering's self-starter.

compared to electrics. At first available only on luxury cars, by 1919, the self-starter was an option on the Ford Model T. (Fig. 2.33)

1910s to 1920s   As a result of more affordable ICE vehicles, the self-starter, improved roads, a rise in interest in longer road trips (known as "touring"), and cheap oil gushing out of the wells in Texas, ICEs totally dominate the roads, temporarily driving electric vehicles to extinction. (Fig. 2.34)

1947   Invented in 1925, the first transistors are built at Bell Labs. Transistors allow for the efficient regulation of power to an electric motor. Previously, resistors had to be used to burn up excess power. Transistors will prove essential in the development of advanced electric vehicles.

1959 to 1960   The Henney Motor Company manufactures 100 of the Henney Kilowatt, an electric car based on the Renault Dauphine. Despite a top speed of 60 mph and a range of 60 mi, there is little demand. (Fig. 2.35)

1964 to 1966   Increasing gas prices revive interest in electric vehicles, and General Motors builds the Electrovair I and II concept cars based on an electric retrofit of the Corvair. Featuring a 115-hp three-phase induction motor powered by a bank of silver-zinc batteries, they reach speeds of up to 80 mph and can go up to 80 mi on a charge. The batteries, however, are expensive and short-lived, wearing out after only 100 charge cycles.

1967 to 1977   American Motors Corporation promotes the Amitron (1967)—later called the Electron (1977)—as an electric concept car, but it is never produced. In 1974, they produce an electric truck, called the Electruck. The U.S. Postal Service purchases 352 of them for use in highly polluted cities.

1968   MIT and Caltech compete in the Great Electric Car Race. Each team departs from its own campus, racing 3,490 mi to the other

FIGURE 2.34 The Lucas gusher at the start of the Texas oil boom.

FIGURE 2.35 The Henney Kilowatt.

FIGURE **2.36** The Apollo 15 lunar rover.

campus. Both teams finish in just over 210 h after 24 h of delays for each (both motors blow out, and MIT's batteries have to be continuously cooled with 200 lb of ice.)

1971 to 1972 The lunar rovers used by the astronauts of Apollo 15, 16, and 17 are electric. Two 36-V silver-zinc potassium hydroxide primary batteries provide each rover with 57 mi of range, almost three times the maximum distance travelled. The real limit to the range of the lunar rovers is the safe walkback distance using life support (in case the rover broke down). Apollo 17 Commander Eugene Cernan sets the unofficial land speed record on the moon: 11.2 mph. (Fig. 2.36)

1974 The first technology-enabled car-sharing program, Witkar, launches with 35 electric cars and 5 parking and charging stations in Amsterdam. (Fig. 2.37)

1974 As oil prices rise, the minimalist electric Sebring-Vanguard CitiCar sees moderate success, selling 4,444 units over five years. It has a range of around 40 mi and a top speed of around 40 mph. (Fig. 2.38)

1985 Based on earlier work by John Goodenough, Stanley Whittingham, Rachid Yazami, and Koichi Mizushima, Akira Yoshino develops a prototype lithium-ion battery. In 1991, Yoshio Nishi leads a team at Sony and Asahi Kasei to develop a commercial lithium-ion battery.

1990 to 2002 In January, General Motors chairman Roger Smith demonstrates an electric concept car, the Impact, at the Los Angeles Auto Show. In April, Smith announces that the Impact will become

FIGURE 2.37 The Witkar.

a production vehicle. Impressed by the potential of the Impact, the California Air Resources Board (CARB) enacts a Zero-Emissions Vehicle (ZEV) mandate, requiring that the seven largest automakers make an increasing portion of their fleet emissions-free—2 percent by 1998, 5 percent by 2001, and 10 percent by 2003—in order to continue selling cars in California. In response to the mandate, these automakers develop their own prototype electric cars. After a two-year pilot program with the Impact, General Motors (GM) releases an updated version in the form of the EV1, featuring lead-acid batteries and a range of 70 to 100 mi. The EV1 is available for lease (only, no purchase) for $399 to $549 per month. In 1999, GM releases a second generation version featuring more efficient nickel-metal hydride (NiMH) batteries with a range of 100 to 140 mi. In 2001, at the urging of GM and other automakers, CARB relaxes its zero-emissions mandate, allowing automakers to focus on hybrid and fuel-cell vehicles. By 2002, a total of 1,117 EV1s have been produced. On February 7, 2002, GM announces that all EV1s will be taken off the road and begins re-possessing them.

FIGURE 2.38   The Sebring-Vanguard CitiCar.

FIGURE 2.39   The General Motors EV1.

Although a small fraction of cars are donated to museums and educational institutions (with their powertrains de-activated), the vast majority are simply crushed. The story of the rise and fall of the EV1 and CARB's zero-emissions mandate is featured in Chris Paine's 2006 documentary, *Who Killed the Electric Car?* (Fig. 2.39)

1997   AC Propulsion demonstrates the tZero, a prototype electric sports car with a 150-kW motor that accelerates from 0 to 60 in 4.07 sec and runs a quarter-mile in 13.24 sec. The original version, using lead-acid batteries, has a range of 80 to 100 mi. (Fig. 2.40)

FIGURE 2.40   The AC Propulsion tZero.

FIGURE 2.41   The Tesla Roadster.

2.5   Twenty-First Century

2003 to 2012   Martin Eberhard commissions AC Propulsion to convert a tZero to run on lithium-ion batteries. The resulting supercar accelerates from 0 to 60 in 3.6 sec and has a range of 300 mi. Eberhard encourages AC Propulsion to produce the car, but they decline. Martin Eberhard and Marc Tarpenning incorporate Tesla Motors to commercialize the tZero themselves. Elon Musk is brought on as an investor, and JB Straubel is hired as CTO. On July 19, 2006, the Tesla Roadster is revealed, with a 0 to 60 time of 3.9 sec and a range of 244 mi, for a base price of $109,000. In February 2008, the first Roadster is delivered. Over the next five years, 2,450 Roadsters are produced and sold. (Fig. 2.41)

2010   There are more than 1 billion cars on the road worldwide.

FIGURE 2.42   The Nissan LEAF.

FIGURE 2.43   The Tesla Model S.

2010   Nissan releases the Nissan LEAF, an electric hatchback. As of 2020, it has a range of 149 mi (twice the range it had originally) and starts at $31,600. Since 2019, a PLUS model has been offered with a range of 226 mi starting at $38,200.[a] (Fig. 2.42)

2012   Tesla releases the Tesla Model S. As of 2020, it has a maximum range of 391 mi (the longest range of any production EV in history, by far), a minimum 0 to 60 time of 2.3 sec (the fastest 0 to 60 time of any production sedan of any kind), and a starting price of $74,990. (Fig. 2.43)

---

[a]For the current generation of EVs, our focus is on vehicles that are already or will soon be available in the United States. Many more EVs are or will soon be available around the world, particularly in China, where there are literally hundreds of companies trying to break into the electric car market.

FIGURE 2.44 The BMW i3.

FIGURE 2.45 The Fiat 500e.

2012 BMW conducts a limited trial of the BMW ActiveE, an electric retrofit of the 1 Series coupe. Per is of 1,100 testers. In 2013, BMW applies the lessons learned from the ActiveE to produce the BMW i3. As of 2020, it has a range of 153 mi and starts at $44,450. (Fig. 2.44)

2013 Fiat releases the Fiat 500e. As of 2020, it has a range of 84 mi and a starting price of $33,460. A new version with a range of 199 mi will be available soon. (Fig. 2.45)

2013 Smart releases the Smart Fortwo EV. As of 2020, it has a range of 58 mi and a starting price of $24,550. Tesla helped develop an earlier version of the Smart EV, and Elon Musk credits the revenue and credibility that came from that partnership as having been essential to the continued success of Tesla. (Fig. 2.46)

FIGURE 2.46   The Smart Fortwo EV.

FIGURE 2.47   The Volkswagen e-Golf.

2014   Volkswagen releases the Volkswagen e-Golf. As of 2020, it has a range of 125 mi and starts at \$31,895, but was quietly cancelled for the 2020 model year in the United States, so its future is uncertain. (Fig. 2.47)

FIGURE 2.48   The Kia Soul EV.

FIGURE 2.49   The Tesla Model X.

2014   Kia releases the Kia Soul EV. As of 2020, it has a range of 243 mi (more than double the range of the previous year's model) and a starting price of $33,950. (Fig. 2.48)

2015   Tesla releases the Tesla Model X. As of 2020, it has a maximum range of 351 mi, a minimum 0 to 60 time of 2.6 sec, and starting price of $79,990. It features an optional Bioweapon Defense Mode which Tesla claims will allow you to "literally survive a military grade bio attack by sitting in your car," while also "clean[ing] the air outside your car, making things better for those around you." (Fig. 2.49)

2016   Chevy releases the Chevy Bolt. As of 2020, it has a range of 259 mi and starts at $37,495. The Bolt has several interesting energy management features such as showing you a range of possible ranges

FIGURE 2.50   The Chevy Bolt.

FIGURE 2.51   The Tesla Model 3.

based on your driving, rather than just a single number, and letting you know how the terrain and outside temperature, as well as your driving technique and your use of climate control, are impacting your efficiency. (Fig. 2.50)

2016   Tesla announces the Model 3. As of 2020, it has a maximum range of 322 mi, a minimum 0 to 60 time of 3.2 sec, and a starting price of $37,990. In 2020, Nick buys one, and uses it to validate our models in Chap. 3. (Fig. 2.51)

2016   Hyundai releases the Hyundai Ioniq Electric. As of 2020, it has a range of 170 mi and a starting price of $33,045. The Ioniq is unique in that it is the first car that can be bought with one of three different drivetrains: hybrid, plug-in hybrid, or all-electric. (Fig. 2.52)

FIGURE 2.52 The Hyundai Ioniq Electric.

FIGURE 2.53 The New Tesla Roadster.

2017 Tesla announces a new version of the Roadster, with a range of 620 mi, a 0 to 60 time of 1.9 sec, and a maximum speed of over 250 mph. The base model costs $200,000. Production is expected to begin in 2021. Assuming the 1.9-sec time holds, it will be the fastest-accelerating production car in the world. (Fig. 2.53)

2018 Jaguar releases the Jaguar I-PACE. As of 2020, it has a range of 234 mi and a starting price of $69,850. Similar to Teslas, the I-PACE gets occasional software updates that improve range and performance, allowing for the cars to actually improve as they age. (Fig. 2.54)

2018 Hyundai releases the Hyundai Kona EV. As of 2020, it has a range of 258 mi and a starting price of $37,190. (Fig. 2.55)

2018 Volkswagen's electric I.D. R. racing car sets a record time of 7 min, 57.148 sec up the Pikes Peak Hill Climb race course. This is the first time a car has traversed the course in less than 8 min, and is an overall record for both gasoline and electric vehicles. (Fig. 2.56)

FIGURE 2.54 The Jaguar I-PACE.

FIGURE 2.55 The Hyundai Kona EV.

FIGURE 2.56 The Volkswagen I.D. R.

FIGURE 2.57 The Tesla Model Y.

FIGURE 2.58 The Audi e-tron.

2019 Tesla announces the Tesla Model Y, with a maximum range of 316 mi and a starting price of $52,990. Deliveries begin in 2020. Although similar to the Model 3 in many ways, the larger Model Y has a more efficient climate control system which uses a heat pump instead of electric resistance heating. (For the importance of this, see Chap. 3.) (Fig. 2.57)

2019 Audi releases the Audi e-tron. As of 2020, it has a range of 204 mi and a starting price of $74,800. Similar to other Audis, the top-of-the-line model features massage seats. (Fig. 2.58)

FIGURE 2.59 The Porsche Taycan.

FIGURE 2.60 The Kia Niro EV.

2019 Porsche releases the Porsche Taycan. As of 2020, it has a range of 201 mi and a starting price of $103,800. At 800 V, the voltage of the Taycan battery is twice the voltage of most other EVs, which translates into a corresponding decrease in the current required for both charging and operation. (Fig. 2.59)

2019 Kia releases the Kia Niro EV. As of 2020, it has a range of 239 mi and starts at $38,500. The Niro is equipped with a Coasting Guide Control system that advises the driver on the optimal time to take their foot off the accelerator in order to maximize regenerative braking and extend range. (Fig. 2.60)

FIGURE 2.61   The Tesla Cybertruck.

2019   Tesla announces the Tesla Cybertruck, with a maximum range of over 500 mi, a minimum 0 to 60 time of less than 2.9 sec, a maximum towing capacity of more than 14,000 lb, and a starting price of $39,900. Production is expected to begin in 2021. (Fig. 2.61)

2020 and Beyond   For what might happen in the future of electric vehicles, see Chap. 9.

2.6   Conclusion

Electric vehicles have had a long and storied history, full of triumphs and tribulations. While it remains to be seen what will happen in the future, one might hope that the newest generation of electric vehicles will finally prevail where previous generations have so far failed.

2.7   Homework Problems

Rather than simply asking you to remember specific dates and milestones in the history of electric vehicles, the following homework problems ask you to read between the lines and analyze why interest in electric cars has waxed and waned over the centuries.

2.1 What key developments enabled the invention of electric vehicles in the nineteenth century?

2.2 In the first generation of commercial automobiles (around the turn of the twentieth century), what initially made electric vehicles attractive compared to ICE vehicles?

2.3 What contributed to the decline of electric vehicles in the early twentieth century?

2.4 What key developments renewed interest in electric vehicles in the mid-twentieth century?

2.5 Why didn't they catch on at the time?

2.6 What key developments enabled the revival of electric vehicles in final decade of the twentieth century?

2.7 What contributed to the decline of electric vehicles in the early years of twenty-first century?

2.8 What key developments enabled the current generation of electric vehicles?

2.9 Based on everything you know about the past, what factors will determine whether the current generation of electric vehicles finally represents a lasting breakthrough, or whether, like previous generations, it will be short-lived?

Notes

For full-length books on electric vehicle history, see:

1. Segrave, Kerry. *The Electric Car in America, 1890–1922: A Social History.* Jefferson, NC: McFarland & Company, Inc., 2019.

2. Burton, Nigel. *A History of Electric Cars.* Ramsbury, England: Crowood, 2013.

3. Schiffer, Michael Brian. *Taking Charge: The Electric Automobile in America.* Washington, DC: Smithsonian Institution Press, 1994.

Image Credits

Figure 2.1: Standard of Ur — ⑤

Figure 2.2: Verbiest's automobile — ⑤

Figure 2.3: Cugnot's automobile — ⓪ Joe deSousa

Figure 2.4: Voltaic pile — ⑤ Alessandro Volta

Figure 2.5: Rivaz engine — ⑤ François Isaac de Rivaz

Figure 2.6: Božek's oil-powered steam car — ⑤ Postrach

Figure 2.7: Ørsted's compass experiment — ⑤ Agustin Privat-Deschanel

Figure 2.8: Ampere's solenoid — ⑤ H. Schellen

Figure 2.9: Faraday's homopolar motor — ⑤ Michael Faraday

Figure 2.10: Sturgeon's electromagnet — ⑤ William Sturgeon

Figure 2.11: Jedlik's DC motor — ⓒⓘⓞ Jedlik Ányos Társaság

Figure 2.12: Faraday's induction experiment — ⑤ Arthur William Poyser

Figure 2.13: Davenport's DC motor — ⑤ Thomas Davenport

Figure 2.14: Daniell cell — ⑤ Gillard

Figure 2.15: Jacobi's electric motor — Ⓢ

Figure 2.16: Galvani — Ⓢ

Figure 2.17: Planté's lead-acid cell — Ⓢ Gaston Planté

Figure 2.18: La Mancelle — ⓒⓘⓞ Buch-t

Figure 2.19: Trouvé electric tricycle — Ⓢ Alexis Clerc

Figure 2.20: Lichterfelde electric tramway — Ⓢ

Figure 2.21: Electromote — Ⓢ

Figure 2.22: Benz Patent-Motorwagen — Ⓢ

Figure 2.23: Tesla induction motor — Ⓢ Nikola Tesla

Figure 2.24: Flocken Elektrowagen — Ⓢ

Figure 2.25: Squirrel cage induction motor — Ⓢ Mikhail Dolivo-Dobrovolsky

Figure 2.26: Columbia Electric Phaeton Mark III — Ⓢ *Century Magazine*

Figure 2.27: Jeantaud Duc — Ⓢ *La Vie au Grand Air*

Figure 2.28: La Jamais Contente — Ⓢ

Figure 2.29: Pieper parallel hybrid — Ⓢ Henri Pieper

Figure 2.30: Lohner-Porsche Mixte Hybrid — Ⓢ

Figure 2.31: Baker Torpedo — Ⓢ from Detroit Public Library

Figure 2.32: Ford Model T — Ⓢ Harry Shipler

Figure 2.33: Reproduced from Kettering's original drawing — Ⓢ from General Motors Archive

Figure 2.35: Henney Kilowatt — Ⓢ DRoberson

Figure 2.36: Lunar Rover — Ⓢ NASA / Dave Scott

Figure 2.37: Witkar — ⓒⓘⓞ Amsterdam Museum

Figure 2.39: General Motors EV1 — ⓒⓘⓞ Rick Rowen

Figure 2.40: AC Propulsion tZero — ⓒⓘⓞ AC Propulsion

Figure 2.41: Tesla Roadster — ⓒ (with unrestricted use) Tesla Motors

Figure 2.42: Nissan LEAF — ⓒⓘⓞ Vauxford

Figure 2.43: Tesla Model S — ⓒⓘⓞ Granada

Figure 2.44: BMW i3 — ⓒⓘⓞ Riley

Figure 2.45: Fiat 500e — ⓒⓘⓞ Mr.choppers

Figure 2.46: Smart Fortwo EV — ⓒⓘⓞ chuckoutrearseats

Figure 2.47: Volkswagen e-Golf — ⓒⓘⓞ Andrzej Otrębski

Figure 2.48: Kia Soul EV — ⓒⓘⓞ Matti Blume

Figure 2.49: Tesla Model X — ⓒⓘⓞ Christopher Ziemnowicz

Figure 2.50: Chevy Bolt — ⓒⓘⓞ Kevauto

Figure 2.51: Tesla Model 3 — ⓒⓘⓞ Vauxford

Figure 2.52: Hyundai Ioniq Electric — ©①◎ M 93

Figure 2.53: New Tesla Roadster — ©①◎ Editor928111

Figure 2.54: Jaguar I-PACE — ©①◎ Vauxford

Figure 2.55: Hyundai Kona EV — ©①◎ Matti Blume

Figure 2.56: Volkswagen I.D. R — ©①◎ Alexander Migl

Figure 2.57: Tesla Model Y — ©①◎ Daniel.Cardenas

Figure 2.58: Audi e-tron — ©①◎ Alexander Migl

Figure 2.59: Porsche Taycan — ©①◎ Alexander Migl

Figure 2.60: Kia Niro EV — ©①◎ Alexander Migl

Figure 2.61: Tesla Cybertruck — ©①◎ Kruzat

CHAPTER **3**

Vehicle Dynamics

3.1 Introduction

In considering a new source of energy for vehicles, it's important to understand how much energy is required to move a vehicle. Therefore, in this chapter, we will derive a simplified model of the dynamics that govern vehicle motion.[1] Given that the vast majority of energy used for transportation (84 percent in the United States) is by vehicles on wheels (cars, trucks, and trains, as opposed to boats and planes), we will focus on this particular kind of vehicle.[2]

3.2 Forces Acting on a Rolling Vehicle

Figure 3.1 shows the major forces at work on a rolling vehicle.

F_{ad} is aerodynamic drag, F_{rr} is rolling resistance, F_{hc} is hill climbing (a proportion of the gravitational force mg as defined by the angle ψ), F_{la} is linear acceleration, $F_{\omega a}$ is angular acceleration, and F_{te}, the sum of all of them, is the total tractive effort required to move the vehicle.

There are a variety of other forces acting on the car, such as lift (the tendency for a fast car to want to take off like an airplane) and wind resistance (above and beyond aerodynamic drag through still air). But the forces illustrated in Fig. 3.1 are sufficient to form a satisfactory model of the vehicle's behavior for our purposes.

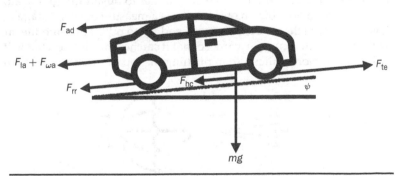

FIGURE 3.1 Force diagram for a rolling vehicle.

Based on Fig. 3.1, the total force (in newtons) required to move a vehicle on wheels is defined by Eq. (3.1):

$$F_{te} = F_{ad} + F_{rr} + F_{hc} + F_{la} + F_{\omega a} \tag{3.1}$$

As shown, the force required to move the car forward, F_{te}, is equal to the sum of all of the forces opposing that forward motion. Each of the terms in Eq. (3.1) will be analyzed in detail in Secs. 3.2.1 to 3.2.6.

3.2.1   Aerodynamic Drag (F_{ad})

One of the primary forces acting on a vehicle is aerodynamic drag, which is the oppositional force imparted on the vehicle by the air that it collides with as it moves forward. The drag force can be calculated using Eq. (3.2):

$$F_{ad} = \frac{1}{2}\rho A C_d v^2 \tag{3.2}$$

where ρ is the density of air (in kg/m^3), A is the frontal area of the vehicle that is pushing against the air (in m^2), C_d is the drag coefficient, and v is the velocity of the vehicle (in m/s).

ρ is a measure of how much mass the vehicle has to push out of the way to travel through it. For example, it's easier to walk through air than it is to walk through water, because air is less dense than water. At sea level, ρ is 1.225 kg/m^3, but as you rise in altitude, this figure decreases. For example, the air density atop Mt. Everest is about a third of what it is at sea level. In general, for our calculations, we will assume our vehicle is driving at sea level.

A is a measure of the area that the vehicle is presenting to the air. If you put the vehicle in front of a backdrop, and looked at it from the front, A is equivalent to the two-dimensional area of the backdrop that would be blocked by the vehicle. The bigger the area, the more air needs to be pushed out of the way to get through. As an analogy, imagine pushing the point of a large umbrella through the air when it's closed as shown in Fig. 3.2. Now imagine pushing the top of the umbrella through the air when it's open: the larger area presented by the open top of the umbrella makes it harder to push through the air.

Aside from how big an object is, C_d is a measure of how "draggy" it is. If you turn the umbrella around and try to pull it through the air with the bottom facing forward, so that it catches the air like a bowl, it will be even more difficult than pushing the top through the air, even

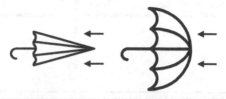

FIGURE 3.2 The second umbrella has larger A.

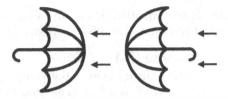

FIGURE 3.3 The second umbrella has larger C_d.

though their areas are exactly the same. This means that the concave bottom of the umbrella has a higher C_d than the convex top of it. (Note that in Figs. 3.2 and 3.3, the arrows represent the relative flow of air as all of the umbrellas move through it to the right.)

Technically, C_d is a function of the circumstances that a vehicle finds itself in, including its velocity. But for simple analyses like the one we're doing here, C_d is usually treated as a constant for a given vehicle.

Given that C_d and A are both constants relating to a particular vehicle, they are often combined to create a compound constant, $C_d A$. The following table shows the values of C_d and $C_d A$ for several popular electric vehicles:

	C_d	$C_d A$
Tesla Model 3	0.23	0.501 m²
Tesla Model S	0.24	0.562 m²
Nissan LEAF	0.29	0.725 m²

Why is F_{ad} proportional to v^2? F_{ad} is the force required for the car to push the air out of the way: in other words, it's the force required to change the momentum of the air in front of the car. Momentum is equal to mass times velocity ($p = mv$), and force is equal to mass times acceleration ($F = ma$), so force is equal to change in momentum over time ($F = m\frac{dv}{dt}$).

Imagine that the front of the car was able to push all of the air in front of it out of the way at velocity v. The amount of the air being pushed out of the way each second is equal to the density of the air, ρ (in kg/m³), times the frontal area A (in m²), times the velocity (in m/s), for final units of kg/s, mass over time. If we multiply by the velocity (m/s) a second time, we get exactly what we're looking for: mass times velocity (i.e., momentum) over time, which gives us the force required to push the air out of the way.

The reason we multiply $\rho A v^2$ by a $C_d < 1$ is that the front of the car isn't actually pushing 100 percent of the air out of the way: a good proportion of it is actually slipping by, so the total force is less than the full $\rho A v^2$. The reason we multiply it all by 1/2 is simply an artifact of the way that C_d is defined elsewhere in the field of fluid dynamics. In an alternate universe, we could've defined all values of C_d as half of their current value and took the 1/2 out of this equation, but for reasons unrelated to this equation, it's more elegant to define C_d as the larger value and add the $1/2$ to this equation.

3.2.2   Rolling Resistance (F_{rr})

In addition to overcoming air resistance, the vehicle needs to overcome rolling resistance, which is a loss generated by the wheels in contact with the road surface. While there are many things that can contribute to rolling resistance, the primary component of it is hysteresis: as the wheel rolls along the road, the wheel (and to a lesser extent, the road surface) is constantly being deformed, which causes a loss of energy in the form of heat.

Rolling resistance is calculated by Eq. (3.3):

$$F_{rr} = \mu_{rr}mg \tag{3.3}$$

where μ_{rr} is the coefficient of rolling resistance, m is the mass of the vehicle (in kg), and g is gravity (9.81 m/s^2). Alternatively, mg can be thought of as the weight of the vehicle (in newtons).

The direct proportionality of rolling resistance to weight makes sense: the harder the vehicle is pressing down on its wheels, the more those wheels are likely to deform, resulting in greater rolling resistance. (As an interesting aside, this is also why it's important to keep your car tires inflated to the recommended pressure: when the pressure is low, the tires deform more, resulting in higher rolling resistance, and as a result, higher energy consumption.)

The direct proportionality of rolling resistance to weight also gives us a helpful way of thinking about what it actually means. If the μ_{rr} of our tires is 0.01 (a common value), that means that the force (in pounds) required to move a load on those tires is 0.01 times (i.e., 1 percent) of the weight of that load in pounds. Therefore, it takes only 1 lb of force to move a load of 100 lb on wheels with a μ_{rr} of 0.01.

The following table shows some common values of μ_{rr}:

	μ_{rr}
Good Car Tire	0.006
Average Car Tire	0.010
Bad Car Tire	0.015
Car Tire on Sand	0.3

And this table shows the value of m for several popular electric vehicles:

	m
Nissan LEAF	1,500 kg
Tesla Model 3	1,611 kg
Tesla Model S	2,200 kg

3.2.3   Hill Climbing (F_{hc})

Aerodynamic drag and rolling resistance are the two major forces opposing a vehicle traveling at a constant velocity on a flat surface, but if you're traveling up a hill, you'll also need to account for gravity. The force required to travel up a hill is calculated by Eq. (3.4):

$$F_{hc} = mg \sin \psi \qquad (3.4)$$

where m is the mass of the vehicle (in kg), g is gravity (8.91 m/s²), and ψ is the vertical angle of the road relative to flat (in radians).

Figure 3.4 illustrates why sine is the appropriate function to use here. The two angles ψ are congruent because they have the same relationship to horizontal and vertical. The force pulling back on the car is $mg \sin \psi$ because we're interested in the opposite side of the vertical triangle relative to the hypotenuse of the vertical triangle, and sine gives us the ratio of opposite to hypotenuse (the SOH of the famous SOH CAH TOA).

In most cases, the hill climbing force equation can be simplified by using the small angle approximation, which states that for small angles (in radians):

$$\sin \theta \approx \theta \qquad (3.5)$$

This fact is clearly illustrated in Fig. 3.5.

The error of the small angle approximation is less than 1 percent until you reach 0.24 rad (14°), so as long as your roads are flatter than that, it's a perfectly reasonable assumption to make. And your roads are almost certainly flatter than that! For reference, the maximum allowable angle of federally funded highways in the United States is around 3.5°, and the maximum recorded angle of any road in the world, Ffordd Pen Llech in Wales, is 20.5°. Even on Ffordd Pen Llech, the error of using the small angle approximation is just over 1 percent, far less than the error that will be introduced by other aspects of our calculations.

So, in general, we're free to simplify to:

$$F_{hc} \approx mg\psi \qquad (3.6)$$

as long as ψ is in radians. In any case, it's important to note that if the vehicle is going downhill, ψ, and therefore F_{hc}, will be negative,

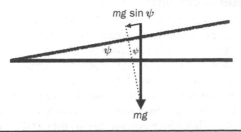

FIGURE 3.4 Hill climbing force is equal to $mg \sin \psi$.

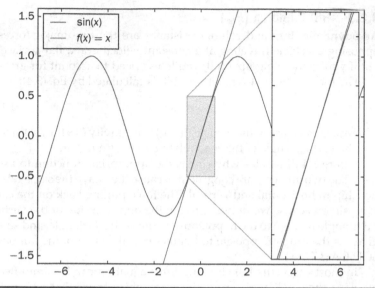

FIGURE 3.5 The small angle approximation.

as gravity pulls the vehicle forward down the slope and reduces the force that the motor needs to provide in order to keep the vehicle moving forward (or increases the force required to slow it down when braking).

3.2.4   Linear Acceleration (F_{la})

If the vehicle is traveling at constant velocity, the three forces described above are sufficient to form a basic model of the vehicle's behavior. But much of the time (and nearly all of the fun times!) a vehicle is also accelerating. The linear acceleration of a vehicle (along the road) is defined by the basic force law:

$$F_{la} = ma \tag{3.7}$$

The faster you accelerate, and the more massive the vehicle, the more force will be required. Note that if the vehicle is slowing down, a, and therefore F_{la}, will be negative.

3.2.5   Angular Acceleration ($F_{\omega a}$)

While it's clear that a force is required to accelerate a vehicle down the road, there's another kind of acceleration going on that's a bit less obvious. Every time the vehicle accelerates, there are a variety of rotational parts inside the vehicle that need to be spun up, which also requires a force (or more accurately, a torque). Given that the force required for angular acceleration within the vehicle is usually much smaller than the force required for linear acceleration of the vehicle itself, and the calculations involved are significantly more difficult, requiring numbers that are much less readily available, we will follow the suggestion of Larminie & Lowry (2012) and use a fudge factor that assumes $F_{\omega a}$ is around 5 percent of F_{la}.[3]

Therefore, we will say that:

$$F_{la} + F_{\omega a} \approx 1.05 F_{la} \tag{3.8}$$

If the vehicle is slowing down, $F_{la} + F_{\omega a}$ will be negative, as a negative force is required to reduce both the linear momentum of the vehicle and the angular momentum of its rotating parts.

3.2.6 Tractive Force (F_{te})

As we saw at the beginning of the chapter, the total tractive effort required to move the vehicle is simply the sum of all of the forces we have discussed:

$$F_{te} = F_{ad} + F_{rr} + F_{hc} + F_{la} + F_{\omega a} \tag{3.9}$$

Elaborating each term as we have defined them, we get:

$$F_{te} = \frac{1}{2}\rho A C_d v^2 + \mu_{rr}mg + mg \sin\psi + 1.05ma \tag{3.10}$$

Putting in our assumptions and approximations for a vehicle driving at sea level on gentle slopes, we get:

$$F_{te} = 0.6125 A C_d v^2 + 9.81\mu_{rr}m + 9.81m\psi + 1.05ma \tag{3.11}$$

To clean things up a bit, let's divide the last three terms by $9.81m$ so we only have to multiply by m once. This leaves us with the following elegant equation for the total force required to move a rolling vehicle on gentle slopes at sea level:

$$F_{te} = 0.6125 A C_d v^2 + 9.81m\left(\mu_{rr} + \psi + 0.107a\right) \tag{3.12}$$

In order to calculate the force required to move a rolling vehicle, all we need to know is: its frontal area (A), its drag coefficient (C_d), its velocity (v), its mass (m), the coefficient of rolling resistance for its wheels (μ_{rr}), the angle of the road (ψ), and the vehicle's current rate of acceleration (a), if any.

3.3 Power Required for Rolling Motion

Understanding the force required to move the vehicle is a good first step, but in the end, what we're really interested in is the power required, as this will allow us to determine the energy that will be drawn from the battery over time.

Because power is work over time, work is force times distance, and velocity is distance over time, power is equal to force times velocity:

$$P_{te} = F_{te}v \tag{3.13}$$

Combining Eqs. (3.12) and (3.13) results in Eq. (3.14) for the power required to move a rolling vehicle on gentle slopes at sea level:

$$P_{te} = 0.6125 A C_d v^3 + 9.81 m v \left(\mu_{rr} + \psi + 0.107a \right) \qquad (3.14)$$

This powerful equation (no pun intended) will form the basis of our entire analysis going forward, so make sure you fully understand it before continuing.

3.3.1 Power Required for Real-World Designs

To truly understand the significance of this Eq. (3.14), let's put in some real-world numbers. Several of these parameters (A, C_d, m, and μ_{rr}) depend on the vehicle we're driving. Conveniently, these parameters have been calculated or measured by vehicle manufacturers, researchers, and enthusiasts, and are readily available for a variety of vehicles. By entering these parameters into Eq. (3.14), we can generate a power equation unique to each vehicle.

For example, let's consider the Tesla Model 3. Tesla has reported that the Model 3's C_d is 0.23, and Curt Austin of Bike Calculator has calculated (with the help of Photoshop), that its frontal area is 2.18 m^2, for a combined $C_d A$ of 0.501 m^2. Assuming the Model 3 is running on average tires with a μ_{rr} of 0.01, and has a mass of 1,673 kg (1,611 kg for the car, plus 62 kg for an average driver), here's the characteristic power equation for the Model 3:

$$P_{\text{Model 3}} = 0.307 v^3 + 16412 v \left(0.01 + \psi + 0.107a \right) \qquad (3.15)$$

The Nissan LEAF is reported to have a C_d of 0.29, a $C_d A$ of 0.725, and a mass of 1,562 kg (including average driver). Given those parameters, here's the characteristic power equation for the LEAF:

$$P_{\text{LEAF}} = 0.444 v^3 + 15323 v \left(0.01 + \psi + 0.107a \right) \qquad (3.16)$$

The slight differences between Eq. (3.15) and Eq. (3.16) account for the fact that the Model 3 is a bit heavier (resulting in a larger second term), and the LEAF is a bit less aerodynamic (resulting in a larger first term). Similar equations can be generated for any vehicle for which A, C_d, m, and μ_{rr} are known.

For example, here's the power equation for one of the least efficient vehicle designs in history, the Hummer H2 (the discontinued gasoline version, not the upcoming Hummer EV):

$$P_{\text{Hummer}} = 1.507 v^3 + 30038 v \left(0.01 + \psi + 0.107a \right) \qquad (3.17)$$

As expected, both terms are significantly larger for the Hummer than they are for either the Model 3 or LEAF.

On the other end of the spectrum, the Aptera 2e was one of the most efficient vehicles ever designed. Unfortunately, the 2e was never

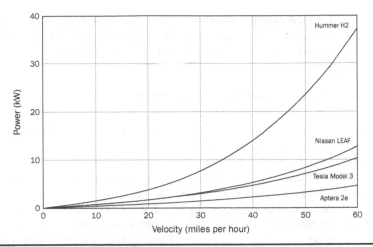

Figure 3.6 Power required for constant velocity.

produced, but it underwent extensive testing and optimization for efficiency, resulting in this lean power equation:

$$P_{\text{Aptera}} = 0.116v^3 + 8652v\,(0.01 + \psi + 0.107a) \qquad (3.18)$$

Clearly there's a large difference in the power required to move a Hummer and an Aptera down the road, but how much of a difference exactly? Figure 3.6 shows the power required to move the four different vehicles we've analyzed down the road at the constant velocity defined by the x-axis. Note that this is the power required to travel at a fixed speed, not the power required to accelerate to that speed—we'll account for acceleration a little later.

As you can see from Fig. 3.6, the difference in power requirements is dramatic. At 60 mph, the Hummer requires more than eight times the power required by the Aptera. As expected, the Model 3 and LEAF both require significantly more power than the Aptera, but significantly less power than the Hummer. Specifically, at 60 mph, the Model 3 requires less than a third of the power required by the Hummer.

In addition to comparing four specific vehicle designs, Fig. 3.6 is interesting because it illustrates what are likely to be the upper and lower bounds for the power required for a vehicle that might reasonably be called a car. (Okay, the Hummer is arguably approaching a military transport, and the Aptera is arguably an enclosed motorcycle, but this just means that our bounds are set conservatively high and low.) Every other kind of car that might reasonably be produced will have a power curve somewhere between these upper and lower bounds. Other types of vehicles—such as trains, buses, RVs, and motorcycles—may have power requirements that are higher or lower, but for a regular car, the power curve will almost certainly fall within this range.

Figure 3.6 is useful for comparing the total power required for different vehicles, but let's take a step back for a moment and see what's going on under the hood. For a given vehicle, how does each force acting on the vehicle factor into the power required to move it?

3.3.2 Power Required for Air Resistance and Rolling Resistance

Consider a Model 3 traveling on flat ground at sea level. Figure 3.7 shows the power required for it to travel at a given speed between 0 and 60 mph. Again, it's important to note that this is the power required to travel at a fixed speed, not the power required to accelerate to that speed—we'll take a look at that soon.

At constant velocity on level ground, there are two components of the power equation: air resistance and rolling resistance. As we would expect to see from Eq. (3.15), Fig. 3.7 shows that rolling resistance is proportional to v, and air resistance is proportional to v^3. Given that the constant associated with rolling resistance (164.12) is much larger than the constant associated with air resistance (0.307), rolling resistance dominates at low speeds (in this case, when $v < 52$ mph), but is overtaken by air resistance at higher speeds (when $v > 52$ mph). The faster we go, the more air resistance dominates, as the power of v^3 eats our lunch.

To understand how powerful air resistance really is, consider a Model 3 driving down Stephen's local freeway, CA-280, with a speed limit of 65 mph, as compared to a Model 3 traveling down Nick's local freeway, TX-130, with a speed limit of 85 mph (the fastest speed limit in the United States). Although 65 mph and 85 mph probably don't seem all that different to the driver, it makes a huge difference to the car. As you can see in Fig. 3.8, it takes more than twice as much power to overcome air resistance at 85 mph as it does to overcome it at 65 mph, even though you're only traveling 30 percent faster!

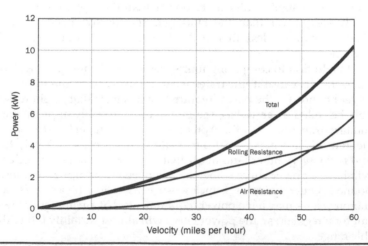

FIGURE 3.7 Power required for a Model 3 to travel at constant velocity.

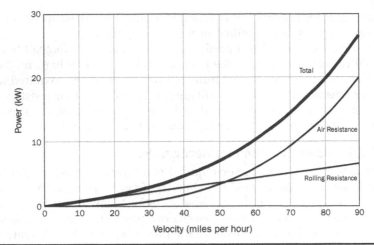

FIGURE 3.8 Power required for a Model 3 to travel at faster velocities.

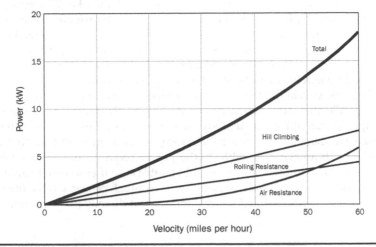

FIGURE 3.9 Power required for a Model 3 to climb a 1° incline.

This is why, if you're ever running low on battery capacity while driving an EV, it's better to slow down (within reason) rather than to speed up. Speaking from personal experience, while driving an EV running on empty, you'll be strongly tempted to speed up in order to "get home faster" to charge, but as Fig. 3.8 shows, the faster you go, the more power you'll use, and the less likely it will be that you'll have enough energy left to make it home!

3.3.3 Power Required for Hill Climbing

The next element of the power equation is hill climbing. Traveling up a hill, how much power does it take to overcome the force of gravity pulling the car backward down the hill? Figure 3.9 shows the power required to move a Model 3 up an incline of 1°. Again, it's important to note that this is the power required to travel at a fixed speed up a slope, not the power required to accelerate up the slope.

How steep is 1°? Not particularly steep. For every football field (100 yd) you travel along the road, you'll have risen only 63 in, the height of an average female human. Even so, the power required to overcome this gentle slope is still greater than either rolling resistance or air resistance, at least at speeds below 60 mph (eventually, as speed increases further, the v^3 of air resistance will once again dominate).

3.3.4   Power Required for Acceleration

At constant velocity, power required is a function of air resistance and rolling resistance (and hill climbing if you're going up a slope). But what about when the car is accelerating? How does the power required to travel at constant velocity compare to the power required to accelerate? Figure 3.10 shows the power required to accelerate a Model 3 at a constant rate of 1 m/s² from 0 to 60 mph on level ground.

How fast is 1 m/s²? Not very fast (okay, to be honest, it's downright slow). At a constant acceleration of 1 m/s², it takes 27 sec to go from 0 to 60 mph. The Model 3 is capable of much faster acceleration, with a top 0-60 time of 3.2 sec. However, as you can see in Fig. 3.10, even a very gentle acceleration of 1 m/s² totally dominates the power equation. This is because the acceleration is directly multiplied by the mass of the vehicle, without being mitigated by a fractional constant like μ_{rr} or ψ as in the case of rolling resistance and hill climbing.

While the contribution of acceleration to the power equation is large, we can at least be thankful that its contribution is linear rather than exponential. Unlike air resistance, which is proportional to the cube of its relevant variable v, the power required to accelerate is directly proportional to its relevant variable a, so if you want to accelerate twice as fast, it only requires twice as much power. On the other hand, the power required to accelerate is also directly proportional to

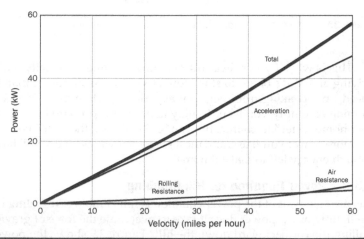

FIGURE 3.10 Power required for a Model 3 to accelerate at 1 m/s².

v, so as you accelerate at a constant rate, the power required to maintain that constant rate of acceleration increases in direct proportion to your speed.

3.4 Modeling Real-World Conditions

The graphs we've drawn so far are interesting, and useful for teasing out the relative importance of each of the forces acting on a rolling vehicle, but they're not particularly useful for analyzing a real-world vehicle, which faces constantly changing v, a, and ψ. In the next section, we'll account for these changing variables, and show how to model a vehicle driving in real-world conditions. We'll start by looking at v and a. Later, we'll add in the effects of ψ.

3.4.1 Modeling *v* and *a* over Time

In modeling v and a that change over time, it would be nice if we had a table that reported the values of v and a over the course of a particular journey. Fortunately, such tables are easy to come by. The most commonly used data for modeling time-variable v and a comes from the "dynamometer driving schedules" published by the U.S. Environmental Protection Agency (EPA). Each driving schedule is published as a table of speeds at 1-sec intervals. While these driving schedules don't necessarily represent any one particular journey, they are designed to be reasonable approximations of a particular kind of journey. For example, the Urban Dynamometer Driving Schedule (UDDS) is designed to represent "city driving conditions."

Figure 3.11 shows a plot of speed over the course of the UDDS.

The acceleration isn't explicitly provided by the schedule, but it can be reasonably approximated for each second by subtracting the current speed from the following one. For example, if we're currently going 10 m/s, and the next second we're going 11 m/s, it's reasonable

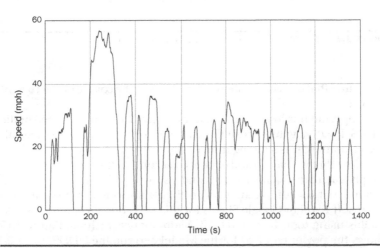

Figure 3.11 The Urban Dynamometer Driving Schedule (UDDS).

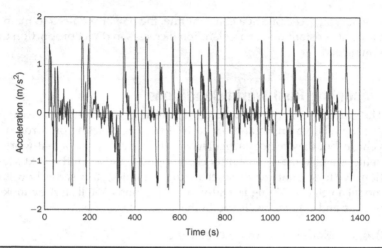

FIGURE 3.12 Acceleration in the UDDS.

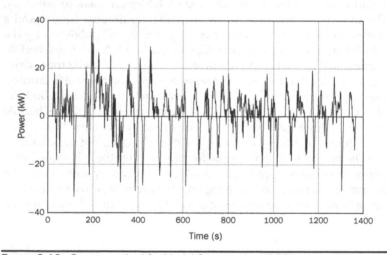

FIGURE 3.13 Power required for Model 3 to run the UDDS.

to assume that our acceleration is 1 m/s² over the course of the current second.

Applying this logic, Fig. 3.12 depicts acceleration over time in the UDDS. Positive values mean the car is speeding up, and negative values mean the car is slowing down.

Now that we have v and a for each second of the schedule, we can simply calculate P_{te} for each second using Eq. (3.15) (for now we'll assume that $\psi = 0$). We are using a simple spreadsheet program like Microsoft Excel or Google Sheets for this task, but it's equally easy to do this using tools like MATLAB, Mathematica, or Julia. Figure 3.13 shows the power required for the Model 3 to run the UDDS.

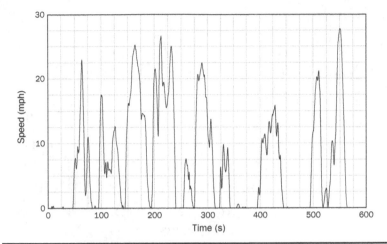

FIGURE 3.14 The New York City Cycle (NYCC).

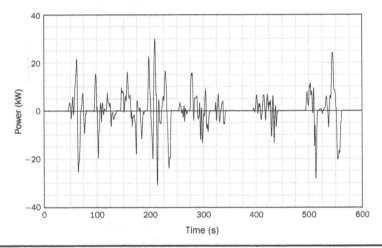

FIGURE 3.15 Power required for Model 3 to run the NYCC.

In this case, positive values mean that power must be expended to move the vehicle, and negative values mean that there is excess power available. This excess power can either be captured by regenerative braking to recharge the batteries (more on this in just a bit), or else it must be dissipated as heat by conventional braking.

The UDDS is just one of many different driving schedules we could use to model the power requirements of the Model 3. Figures 3.14 to 3.17 show the speed curves for a variety of other driving schedules as well as the power required for the Model 3 to run them. The New York City Cycle (NYCC) is designed to model low-speed stop-and-go traffic in the city, and the US06, also known as the Supplemental Federal Test Procedure, is designed to model more aggressive driving at higher speed.

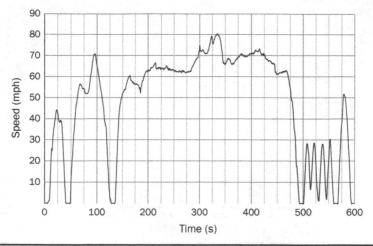

FIGURE 3.16 The US06 Supplemental Federal Test Procedure.

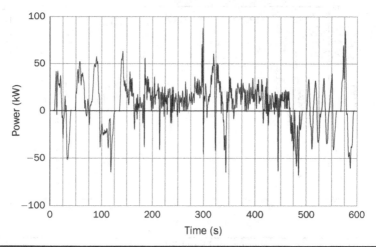

FIGURE 3.17 Power required for Model 3 to run the US06.

Once you know how to generate these kinds of power curves, it's trivially easy to create them for any vehicle for which you know A, C_d, m, and μ_{rr}, and any driving schedule which gives us v as a function of time. The homework problems at the end of this chapter will have you do just that.

3.4.2 Modeling ψ over Time

Now that we've got a handle on modeling v and a over time, let's add in the effect of ψ. While the EPA driving schedules don't account for incline, it's easy to generate a reasonable dataset by using one of the many apps that record motion, such as the Gauges app for iOS. Gauges allows us not only to record our v over time, but also our

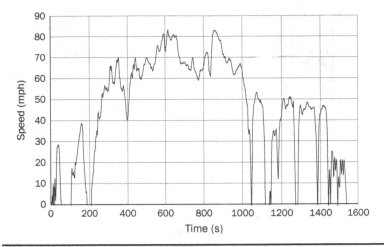

FIGURE 3.18 Speed on Nick's drive from UT.

elevation based on GPS measurements, which will allow us to approximate ψ over time.[a]

Figure 3.18 shows the speed data that Gauges recorded on a recent drive from Nick's office at the University of Texas (UT) to his home in his Tesla Model 3 Long Range. (Note that this wasn't during rush hour, so the roads were relatively clear, and the roads on his route have high speed limits, so he was only speeding by 3 mph over a posted speed limit of 80 mph.)

Figure 3.19 shows the elevation of the route he took from UT to his home and proves that central Texas is indeed quite flat.

While Figure 3.19 doesn't directly give us ψ, it's easy to calculate it from the underlying data. If we subtract the current elevation from the following one, and divide by the time in between the data points (in this case, the time step was 1 sec), we can calculate our effective vertical velocity.

$$v_{\text{vertical}} = \frac{\text{change in elevation}}{\text{time interval}} \tag{3.19}$$

Then, to find the angle of the road, ψ, we simply need to divide this vertical velocity (the side of the triangle opposite the angle we're interested in) by our translational velocity along the road (the hypotenuse of that triangle), and take the arcsin of that ratio.

$$\psi = \arcsin\left(\frac{v_{\text{vertical}}}{v_{\text{translational}}}\right) \tag{3.20}$$

---

[a]It also allows us to record a, but that a includes acceleration in all directions, so for our purposes, it will be more effective to calculate a based on change in v, as we previously did for the driving schedules.

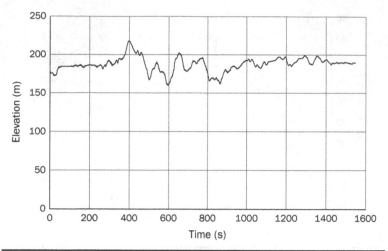

FIGURE 3.19 Elevation on Nick's drive from UT.

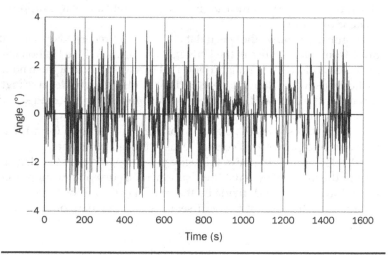

FIGURE 3.20 ψ on Nick's drive from UT.

If we do this for every point in the data set, calculating the angle from the current point to the next point, we can plot the value of ψ over the whole trip.

In calculating the data for Fig. 3.20, there were some outliers that needed to be removed, including one that suggested the angle of the road was 28°, greater than the incline of any road in the world. The cut-off point for reasonable data was set at 3.5°, the maximum allowable angle of federally funded U.S. highways. Even so, there's still quite a bit of noise in the data, but in general, given that there's roughly equal amounts of noise in both directions, it should pretty much cancel itself out.

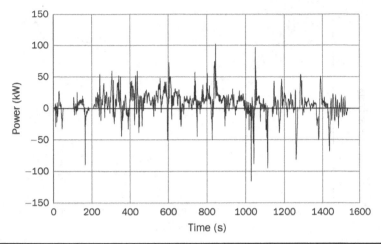

FIGURE 3.21 Power required for Nick's drive from UT.

Now that we have v, a, and ψ for each time step, we can use this data to calculate the total power required at each time step (Fig. 3.21).

3.4.3 Modeling Energy, Efficiency, and Range

So far, we've developed a model for how much power it takes to drive an electric car on a particular route at a particular speed. This is great for building our theoretical understanding of the power required to move a vehicle, but it's still a step removed from what we really care about, which is how far we can drive before the battery runs out. In order to figure this out, we need to calculate several other variables as a function of time: (1) distance traveled, and (2) energy consumed.

To calculate distance traveled, we can simply integrate our velocity function with respect to time. In a computational sense, this means looking at the velocity in each second and multiplying it by that one second to get the distance travelled over the course of that second (meters per second times one second equals meters). Applying this method to the velocity data from Nick's drive from the UT, the graph of distance traveled over time looks like as shown in Fig. 3.22.

As we would expect, the cumulative distance traveled always increases, but the rate at which it increases varies based on how fast the vehicle is traveling (the greater the slope, the higher the speed).

To calculate energy consumption, we can use a similar method, as long as we're careful to consider several important complications. When the power demands are positive (i.e., the motor is working to move the vehicle forward), we can integrate it with respect to time to find the energy required (kilowatts times hours equals kilowatt-hours). However, if we simply integrate the power demands we calculated above, this implies that the drivetrain is 100 percent efficient, which isn't actually the case. In order to arrive at a realistic estimate for the actual energy drawn from the battery, we need to account for this

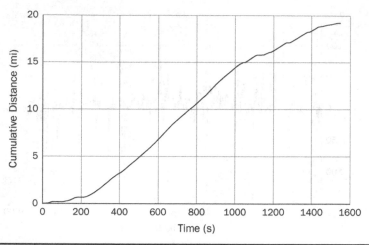

FIGURE 3.22   Distance traveled on Nick's drive from UT.

efficiency. We can do this by dividing the theoretical power required by the efficiency. As a rough estimate, let's suppose the drivetrain is 90 percent efficient at converting energy in the battery into motion in the wheels. This means we should divide the power required at the wheels in every second by 0.9 to arrive at the actual power drawn from the battery every second.

How about when the power demands are negative? In conventional vehicles, this negative power would be provided by the brakes, which would use up the excess energy by heating the brake pads through friction. Fortunately, in most electric vehicles, we don't simply burn up this energy in the brake pads. Instead, a regenerative braking system is employed. In a regenerative braking system, the negative power required for braking is used to run the motor backward, turning it into a generator and recharging the battery (more on exactly how this works in Chap. 4). Of course, this process also has inefficiencies, and not all of the excess energy can be returned to the battery. As a rough estimate, let's suppose that regenerative braking is 70 percent efficient, that is, 70 percent of the negative power available from slowing the vehicle makes it back into the battery. This means that when the power is negative, we will multiply it by 0.7, while keeping its sign negative, which means power is flowing back into the battery.

There's another complication we need to account for. On his drive from UT, Nick was running the A/C and listening to music. Therefore, in addition to consuming energy to transport him, the car was also consuming energy to keep him comfortable and entertained. While the vast majority of energy consumed by an electric vehicle is that which is used to move the vehicle, it's important to consider the other features of the car that consume energy: the headlights, the entertainment system, the A/C, the heater, etc. Taken as a whole, these additional electrical loads are called hotel loads. While this may seem like an odd name for something in a vehicle, they're called hotel loads

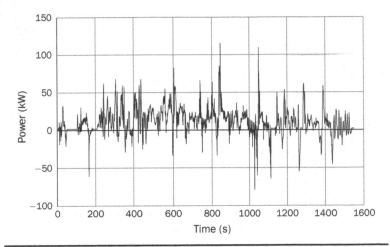

FIGURE 3.23 Actual power required for Nick's drive from UT.

because they're generally related to safe and comfortable human occupancy of the vehicle, rather than its essential function. Depending on which features are being used, they can have either a small or a very large impact on energy consumption and vehicle range.

Thanks to a tip from another driver on the Tesla forums, Nick was able to determine that the power required to keep the car at 72° on the day of his drive while rocking out to his favorite music was 2 kW. (For reference, he also determined that the A/C in the Model 3 draws 4 kW when running at maximum power, and the heater uses 8 kW when running at maximum power. Thankfully, the new Model Y will require significantly less energy for heating because it uses a heat pump instead of electric resistance heating.[b])

To include this hotel load in our calculations, we simply need to add it to every second of our power data, as this load is active and positive regardless of whether the power required to move the vehicle is positive or negative.

Figure 3.23 shows the power entering and exiting the battery after accounting for drivetrain efficiency, regenerative braking, and hotel loads.

Comparing Fig. 3.23 to Fig. 3.21, it looks very similar, except the positive power peaks are a little bit bigger (due to drivetrain efficiency), the negative power peaks are a little bit smaller (due to regenerative braking efficiency), and everything is shifted upward by 2 kW (due to the hotel load).

---

[b]The method for determining this involves turning on the hotel loads in question while the car is fully charged but connected to a charging station. After a few seconds, the charging station will start supplying the electricity required to run those loads, and the amount of power flowing can be read out from the car's energy management screen.

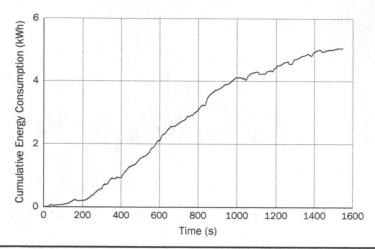

FIGURE 3.24   Energy consumed on Nick's drive from UT.

If we integrate with respect to time, we can graph the cumulative energy consumption, as shown in Fig. 3.24.

Similar to the cumulative distance traveled, the overall trend is up and to the right, but we can see that in moments of braking, the cumulative energy consumption temporarily dips as energy is recovered by the battery.

This is all great in theory, but how does it actually compare to the energy Nick's Model 3 consumed in the real world? Conveniently, the Model 3 provides real-time data about how much energy it consumes. Specifically, in "Trip" mode, the Model 3 graphs battery charge (as a percentage) on the y-axis, and distance (in miles) on the x-axis. If we plot our calculated figures using the same variables, here's how it compares to the graph generated by the car.[c]

Despite the fact that our calculations rely on a variety of rough estimates and assumptions, the overall result is surprisingly accurate. In the end, our model predicted that we would end up with a final battery charge of 67.4 percent, which is exactly what the car reported. And while the two lines in Fig. 3.25 don't always line up exactly, they are within 1 percent of each other at all times.[d] For a simplified model based on basic physics, this is a truly phenomenal result.

---

[c]The graph from the car was reproduced by taking a high quality photograph of the graph and stripping the data out of that photograph using Ankit Rohatgi's amazingly useful WebPlotDigitizer. The calculated graph was generated by subtracting cumulative energy consumption at each second from the initial battery charge reported by the car, then plotting the points at regular distance intervals, using the same granularity (0.1 mi) as was used in the reproduction of the car graph.

[d]Note that the y-axis of Fig. 3.25 is extremely zoomed in, showing only 64 percent to 76 percent on the scale of battery charge from 0 percent to 100 percent, so the small differences that exist between the two lines are magnified.

If we divide the total energy consumed (5.05 kWh or 5,050 Wh) by the total distance traveled (19.2 mi), we see that the average energy consumption per mile (a common measure of electric vehicle efficiency) of Nick's driving was 263 Wh/mi.

$$\text{EV efficiency} = \frac{\text{energy consumed}}{\text{distance traveled}} = \frac{5050 \text{ Wh}}{19.2 \text{ mi}} = 263 \text{ Wh/mi} \quad (3.21)$$

The EPA reports the efficiency of electric vehicles in terms of kWh/100 mi.[4] If we convert to these units, the average energy consumption of Nick's Model 3 Long Range was 26.3 kWh/100 mi, or 26 kWh/100 mi if we round it down to two significant figures in order to match the EPA figures. Coincidentally, this is exactly what the EPA reports for the average energy consumption of the 2020 Model 3 Long Range. We wouldn't necessarily expect a random route on freeways with a speed limit of 80 mph to match the EPA figure exactly, given that those figures are meant to be an average over many different driving conditions, not necessarily representative of any one particular trip, but we're glad to see that once again, our model checks out.

This means that if Nick's Model 3 Long Range had a fully charged 75-kWh battery, it could travel 285 mi on routes that are roughly similar to this one.

$$\text{range} = \frac{\text{battery capacity}}{\text{efficiency}} = \frac{75 \text{ kWh}}{26 \text{ kWh/100 mi}} = 285 \text{ mi} \quad (3.22)$$

While this is less than the 322 mi reported for the Model 3 Long Range, that makes sense given the relatively high speed limits on this particular route (see Fig. 3.8). Assuming the energy required for Nick's drive to UT is similar to the energy required for his trip from UT, this

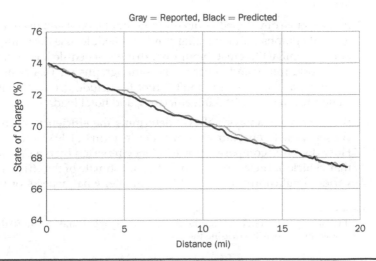

Figure 3.25 Predicted vs. reported state of charge.

means he could make the round-trip commute of roughly 40-mi seven times before he needed to charge.[e]

3.5   Conclusion

In this chapter, we've learned how to model the basic dynamics of a wheeled vehicle in order to estimate the power required for the vehicle to follow a specific driving schedule (with variable v and a) over a specific course (with variable ψ).

Using these estimates, we were able to calculate the energy that would be consumed by a Tesla Model 3 on a given trip. Even though our model was relatively simple and relied on a variety of rough estimates and assumptions, when we compared its predictions to real-world data taken from an actual Model 3, they were within 1 percent of reality. This goes to show just how powerful an understanding of physics can be.

3.6   Homework Problems

3.1 Generate the characteristic force and power equations for an electric vehicle of your choice. (*Hint:* First you'll need to find reasonable estimates for A, C_d, m, and μ_{rr}.)

3.2 Using the power equation you generated above, along with a driving schedule and computational tool of your choice, generate graphs for speed, acceleration, power, distance, and energy consumption over time. A variety of driving schedules can be found here: https://www.epa.gov/vehicle-and-fuel-emissions-testing/dynamometer-drive-schedules (*Hint:* Don't forget to account for drivetrain efficiency, regenerative braking efficiency, and hotel loads!)

3.3 After you've developed your computational model and used it to generate graphs for an existing driving schedule, use the Gauges app (or similar) to create your own driving schedule based on a real-world route, and generate the same graphs for this custom route. (*Hint:* Once again, don't forget to include drivetrain efficiency, regenerative braking efficiency, and hotel loads!)

3.4 Based on your analysis above, determine the efficiency of your chosen vehicle on your chosen route in terms of kWh/100 mi. How does this compare to the efficiency reported by the EPA for that vehicle at fueleconomy.gov? If your estimate of the efficiency varies significantly from the EPA estimate, what aspects of the

[e]In reality, like a good Model 3 owner, he charges it every night, and only to 80 percent to avoid unnecessarily stressing the battery.

route (or perhaps your analysis) do you think might be contributing to this disparity? (As noted at the end of our Model 3 analysis, the existence of a disparity isn't necessarily a problem, as long as you can reasonably explain why it might exist.)

3.5 Assuming that it is driven on routes similar to the one you've analyzed, how far could your chosen vehicle travel with a fully-charged battery? How does this compare to the range that the EPA reports for this vehicle at fueleconomy.gov? Once again, if your estimate varies significantly from the EPA estimate, what aspects of the route (or perhaps your analysis) do you think might be contributing to this disparity?

Notes

1. This chapter builds on analytic frameworks presented in: Larminie, James and John Lowry. *Electric Vehicle Technology Explained.* Hoboken, NJ: Wiley, 2012, and Masias, Alvaro. "Lithium-Ion Battery Design for Transportation." In *Behaviour of Lithium-Ion Batteries in Electric Vehicles: Battery Health, Performance, Safety, and Cost*, edited by Gianfranco Pistoia and Boryann Liaw, 1–34, Cham, Switzerland: Springer, 2018.

2. Davis, Stacy C. and Robert G. Boundy. "Table 2.9. Highway Transportation Energy Consumption by Mode, 1970–2017." *Transportation Energy Data Book: Edition 38.* Oak Ridge, TN: Oak Ridge National Laboratories, 2020. https://tedb.ornl.gov/data/

3. Larminie, James and John Lowry. *Electric Vehicle Technology Explained.* Hoboken, NJ: Wiley, 2012.

4. EPA. "Fuel Economy Guide." http://www.fueleconomy.gov/feg/findacar.shtml

Image Credits

Figure 3.1: Nick Enge, car by monkik on Flaticon

Figure 3.2: Nick Enge, umbrellas by Freepik and Icongeek26 on Flaticon

Figure 3.3: Nick Enge, umbrellas by Freepik on Flaticon

Figure 3.4: Nick Enge

Figure 3.5: ⓞ Stephan Kulla

Figures 3.6 to 3.25: Nick Enge

CHAPTER 4

Motors

4.1 Introduction

In order for an electric vehicle to function, it needs a way to turn the electricity supplied by the battery into the rotational motion required to move the vehicle forward. Fortunately, as we will learn in this chapter, there are many elegant ways of doing this by using a variety of different kinds of electric motors.

At the beginning of this chapter, we'll explore the basic electromagnetic laws that enable motors to function, including the Lorentz Force Law, Coulomb's Law, and Biot-Savart's Law. Then we will proceed to learn about a variety of different types of motors, including brushed DC motors, brushless DC and AC motors (a.k.a. synchronous motors), reluctance motors, and induction motors. In order to explain induction motors, we will also need to understand Faraday's Law, which will further influence our understanding of the other types of motors, as well as enabling them to run backward and generate energy, allowing for regenerative braking.

But before we get ahead of ourselves, let's put our feet on a firm foundation of physics and study the fundamental laws governing electric motors.

4.2 The Lorentz Force Law

The first fundamental law that motors rely on is the Lorentz Force Law. This law states that when a charged particle, like an electron, enters an electric or magnetic field, a force is imparted upon it by those fields. In 1895, Dutch physicist Hendrik Lorentz formulated Eq. (4.1) which elegantly describes exactly how much force is imparted:

$$\overline{F} = q(\overline{E} + \overline{v} \times \overline{B}) \tag{4.1}$$

In Eq. (4.1), $\overline{F}$ is the force imparted, q is the amount of charge, $\overline{E}$ is the electric field, $\overline{v}$ is the velocity of the particle, and $\overline{B}$ is the magnetic field. The bars over $\overline{F}$, $\overline{E}$, $\overline{v}$, and $\overline{B}$ denote that they are vectors, meaning that they have both magnitude and direction, and the direction matters. q, on the other hand, is a scalar quantity, with magnitude only, no direction. In SI units, the force is in newtons, the charge is in coulombs, the electric field is in newtons/coulomb, the velocity is in m/s, and the magnetic field is in teslas, where a tesla is equal to a newton / (coulomb m/s).

As shown in Eq. (4.1), the force imparted on a charged particle by an electric field is equal to q times $\overline{E}$. Because q is a scalar, and includes no directional component, the direction of the force, $\overline{F}$, is the same as the direction of the electric field, $\overline{E}$. Given the simplicity of this relationship, we will leave the electric field component of the equation alone for now, and give greater consideration to the magnetic field component, which is the part most relevant to our study of motors.

The force imparted on a charged particle by a magnetic field is equal to $q(\overline{v} \times \overline{B})$. This makes things a bit more complicated, as now the force depends not only on the magnitude of the charge and the magnitude and direction of the field, but also on the magnitude and direction of the velocity of the particle.

$\overline{v} \times \overline{B}$ is the cross-product of the velocity vector and the magnetic field vector. By definition, this means that the direction of the resulting force vector, $\overline{F}$ will be perpendicular to both the velocity and the magnetic field, as defined by the famous right-hand rule shown in Fig. 4.1.

If you point the index finger of your right hand in the direction of the velocity, and your middle finger in the direction of the magnetic field, your thumb will be pointing in the direction of the force. (In the rare case where the two vectors are parallel, the force is equal to zero.)

Following this rule, Fig. 4.2 shows what happens when a charged particle enters a uniform magnetic field. The particle, with charge q and velocity $\overline{v}$, enters the magnetic field, $\overline{B}$, which, as indicated by the circled dots, is coming straight out of the page. As it enters the field, it is acted upon by the force of $q(\overline{v} \times \overline{B})$, which is perpendicular to both the field and the velocity, causing it to turn to the right (you can check this yourself using the right-hand rule). As it continues to travel, it continues to turn right, so its path through the field becomes a circle.

Of course, in the context of a motor, we aren't actually dealing with solo charged particles. Instead, we're dealing with an electric current, that is, a bunch of charged particles traveling down a wire. Therefore,

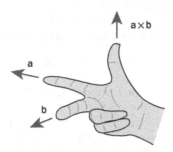

FIGURE 4.1 The right-hand rule.

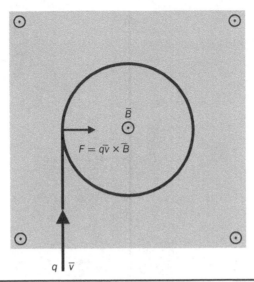

Figure 4.2 Trajectory of a charged particle in a uniform magnetic field.

it will be important to consider how the Lorentz Force Law applies to a wire.

Consider a wire that crosses a magnetic field, as illustrated in Fig. 4.3. The wire is carrying charge at a density of λ coulombs/m, and the individual charges are traveling down the wire at velocity $\overline{v}$.

Consider a small length of wire $d\ell$. The amount of charge in this length of wire is simply the length times the charge density, that is, $\lambda d\ell$. Therefore, according to the Lorentz Force Law [Eq. (4.1)], the force on this small length of wire due to the magnetic field is:

$$\overline{F} = \lambda d\ell \overline{v} \times \overline{B} \tag{4.2}$$

Setting aside $d\ell$ for a moment, we can express $\lambda \overline{v}$ a different way: coulombs per meter times meters per second equals coulombs per second, which is the current, $\overline{I}$, running through the wire, in amperes. Therefore:

$$\overline{F} = \overline{I} d\ell \times \overline{B} \tag{4.3}$$

If we integrate that along the entire length of wire passing through the magnetic field, we get:

$$\overline{F} = \overline{I}\ell \times \overline{B} \tag{4.4}$$

Using the right-hand rule, we can see that in this case, the entire wire will be pulled to the right (Fig. 4.3).

The Lorentz Force Law, which defines the forces exerted on charges traveling through electric and magnetic fields, is the first essential piece of the puzzle of understanding motors. The next pieces are Coulomb's Law and Biot-Savart's Law, which describe, in turn, how electric and magnetic fields are generated.

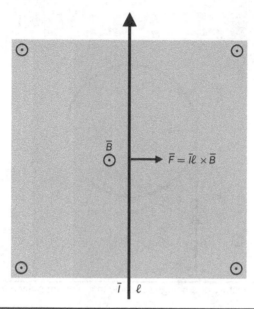

$$\overline{F} = \overline{I}\ell \times \overline{B}$$

FIGURE 4.3 Force on a wire passing through a uniform magnetic field.

4.3 Coulomb's Law

In its most basic form, Coulomb's Law describes the force that two charged objects exert on each other.

$$\overline{F} = \frac{1}{4\pi\epsilon_0} \frac{q_1 q_2 \hat{r}}{r^2} \tag{4.5}$$

As you can see in Eq. (4.5), the force is proportional to the product of the charges $q_1 q_2$, so the larger the charges, the stronger the force. It is also inversely proportional to the square of the distance between them, r^2, so the farther away they are, the weaker the force. The direction of the force vector, $\overline{F}$, is given by the unit vector, $\hat{r}$, which points from one charge to the other and has a value of 1. If the charges are the same (e.g., both negative), they repel, so $\overline{F}$ is negative, and if the charges are opposite, they attract, so $\overline{F}$ is positive. ϵ_0 is the "vacuum electric permittivity," also known as the "electric constant," which quantifies how well an electric field permeates through a vacuum. Taken together, the fractional first term in Eq. (4.5) is known as Coulomb's constant.

The electric field generated by a charged object is expressed as the force per unit charge, newtons per coulomb, or N/C. Therefore, if we divide both sides of Eq. (4.5) by q_1, we get Eq. (4.6), which shows the electric field, $\overline{E}$, generated by q_2:

$$\overline{E} = \frac{\overline{F}}{q_1} = \frac{1}{4\pi\epsilon_0} \frac{q_2 \hat{r}}{r^2} \tag{4.6}$$

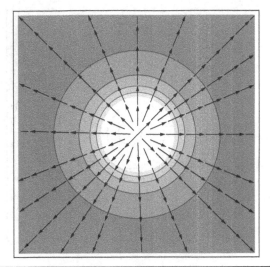

FIGURE 4.4 Electric field surrounding a point charge.

To make it applicable to more than just a pair of charges, we can rewrite this equation as the differential electric field, $d\overline{E}$, due to a differential charge, dq:

$$d\overline{E} = \frac{1}{4\pi\epsilon_0} \frac{dq\hat{r}}{r^2} \tag{4.7}$$

What does this electric field actually look like? Figure 4.4 shows the shape and strength of the electric field surrounding a small parcel of charge, dq. The arrows show the direction of the force acting on a charged object with the same kind of charge (if the charges were opposite, the arrows would point inward). The shades of gray in Fig. 4.4 correspond to the strength of the field: the lighter it is, the stronger the field, and the darker it is, the weaker the field. Note that Fig. 4.4 only illustrates the field in the two-dimensional plane of the page: in actuality, the field surrounds dq in all directions, and the circles in the diagram actually represent concentric spheres surrounding the charge.

If we want to consider several (or many) charges, the resulting field is simply the sum of all of their individual fields:

$$\overline{E} = \frac{1}{4\pi\epsilon_0} \sum_i \frac{dq_i\hat{r}_i}{r_i^2} \tag{4.8}$$

where dq_i are the differential charges indexed by the sum, and $\overline{E}$ is the total field due to all of the charges.

4.4 Biot-Savart's Law

In Sec. 4.2, we saw that a wire carrying current through a magnetic field feels a force from that field. Interestingly, in addition to being

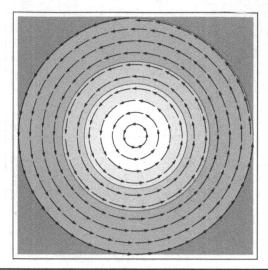

FIGURE 4.5   Magnetic field surrounding a current-carrying wire.

acted on by magnetic fields, current-carrying wires also generate their own magnetic fields. Biot-Savart's Law describes this phenomenon:

$$d\overline{B} = \frac{\mu_0 I}{4\pi} \frac{d\overline{\ell} \times \hat{r}}{r^2} \tag{4.9}$$

Notice the similarity to Coulomb's Law [as expressed in Eq. (4.7)]. In this case, $d\overline{B}$ is the differential magnetic field due to the current I flowing in the short segment of wire $d\overline{\ell}$. The vector $\hat{r}$ is the unit vector pointing from that differential length to the test point, and μ_0 is the "permeability of free space," also known as the "magnetic constant," which quantifies how well a magnetic field permeates through a vacuum. Recall from Sec. 4.2 that the units for magnetic field are teslas, or newtons / (coulomb m/s).

Consider a small length of wire $d\overline{\ell}$ carrying current I through the page, toward you. Figure 4.5 shows the resulting magnetic field in the plane of the page.

Why are the field lines circular? According to Biot-Savart's Law, the direction of the magnetic field vector, $d\overline{B}$, is determined by the cross-product $d\overline{\ell}$, the length of wire being considered, and $\hat{r}$, the unit vector pointing from that length of wire to our test point. Consider a test point to the right of the wire in Fig. 4.5. If you point the index finger of your right hand toward you, along the wire, and your middle finger to the right toward the test point, your thumb points up, just like the arrows in Fig. 4.5. If you re-position the test point by moving your right hand around in space, always pointing your index finger toward you, and your middle finger toward the re-positioned test point, you'll see that the magnetic field is indeed circular.

While the original right-hand rule will never steer you wrong, if you want an easier (and more wrist-friendly) way of determining the

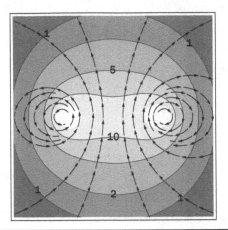

Figure 4.6   Magnetic field surrounding a single loop of wire.

magnetic field lines around a wire, there is a second right-hand rule that's much simpler. Point your thumb in the direction of the current, in this case, toward you, then curl your other four fingers around the wire. The direction of your fingers is the direction of the field lines.

The magnetic field surrounding a longer wire can be calculated by simply summing up the differential magnetic fields due to the differential lengths of wire:

$$\overline{B} = \frac{\mu_0 I}{4\pi} \sum_i \frac{d\overline{\ell}_i \times \hat{r}_i}{r_i^2} \tag{4.10}$$

So far, we've seen that current traveling along a straight wire generates a circular magnetic field. In the context of motors, it will also be useful to understand the kind of magnetic field that is generated by a circular wire.

Consider a single loop of wire through which current travels out of the page on the left and into the page on the right. Based on Eq. (4.10), Fig. 4.6 shows the magnetic field generated by this loop of wire, as well as the relative strength of the field at various points surrounding it (represented by both shades of gray and relative numerical strength values). Take a moment to check that the direction of the field lines makes sense using our new right-hand rule.

If you're having trouble visualizing the wire in Fig. 4.6, imagine that you've cut a donut in half from the top, and you're looking at the ends of the two cut arms of the donut at eye level. Current is flowing toward you out of the left arm of the donut, and away from you into the right arm of the donut.

Interestingly, there's a third right-hand rule that can help us quickly identify which way the magnetic field will flow through a loop of wire. If you curl your fingers in the direction of the current through the loop (i.e., into the page on the right and out of the page on the left), your thumb will point in the direction of the field!

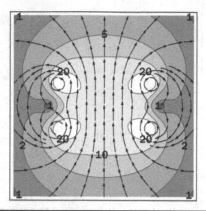

FIGURE 4.7   Magnetic field surrounding two loops of wire.

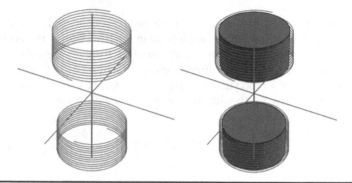

FIGURE 4.8   Helmholtz coils with and without an iron core.

Now, let's consider two parallel loops of wire stacked on top of each other (in other words, we've now stacked two half donuts on top of each other). Once again, the current is traveling out of the page on the left and into the page on the right. Figure 4.7 shows the direction and relative strength of the magnetic field surrounding these loops.

Note that this double loop of wire fills the entire interior volume with a nearly uniform magnetic field. Whereas a linear wire generates a circular magnetic field, circular wires can be used to generate a linear magnetic field!

A structure like this, with two parallel loops of wire carrying current in the same direction, is called a Helmholtz coil. In practice, each of the two loops will actually consist of many windings of wire, rather than just a single loop. This is illustrated on the left side of Fig. 4.8.

If there are N windings, the magnetic field inside the Helmholtz coil becomes N times stronger. If we want to make the field even stronger yet, we can add an iron core inside each of the sets of windings, as illustrated on the right side of Fig. 4.8. Because iron conducts a magnetic field better than a vacuum, this means we also get to multiply the strength of the field by the relative permeability of iron, k, which can be several hundred for the ferrite cores used in motors.

Equation (4.11) shows the final result (yes, it's simply Eq. (4.10) multiplied by N and k):

$$\bar{B} = \frac{k\mu_0 NI}{4\pi} \sum_i \frac{d\bar{\ell}_i \times \hat{r}_i}{r_i^2} \qquad (4.11)$$

4.5   Brushed DC Motors

We now have the background required to understand the so-called brushed direct current (DC) motor, first demonstrated by Hungarian priest Ányos Jedlik in 1828 and still in use today. Figure 4.9 tells the tale of a brushed DC motor.

At the top and bottom of the diagram, we see a Helmholtz coil with iron cores, as developed in the previous section. These two magnetic poles are the main component of the so-called stator, which is the stationary part of the motor. In between the two poles of the stator, we see a rectangular loop of wire that represents the rotor, which is the part of the motor which rotates. But why and how does the rotor rotate?

When the rotor is in the position shown in Fig. 4.9, the current is flowing through the rectangular loop of wire in the direction indicated by the arrows I_1 and I_2. Each of these currents interacts with the magnetic field created by the stator (represented by the dashed lines) in the way we described back in Sec. 4.2.

According to the Lorentz Force Law, a wire passing through a uniform magnetic field experiences a force in the direction of $\bar{I} \times \bar{B}$. The

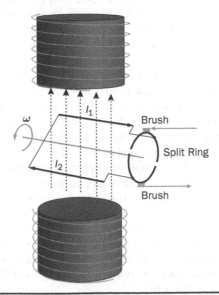

FIGURE 4.9   Operation of a brushed DC motor.

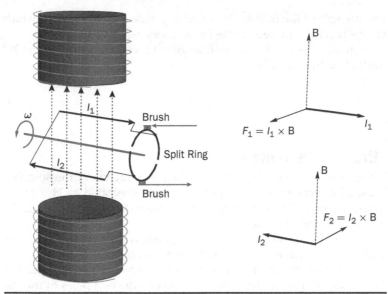

Figure 4.10 Vector analysis of a brushed DC motor.

direction of these vectors and the resulting force vectors for each part of the loop are illustrated in Fig. 4.10.

Using the original, right-angle right-hand rule, if we point our index finger in the direction of $\overline{I}$ (back over our right shoulder) and our middle finger in the direction of $\overline{B}$ (straight up), we can see that the top wire experiences a force that pulls it out of the page to the left, and the bottom wire experiences a force that pushes it into the page to the right. Working together, these two forces create a counterclockwise torque, spinning the rotor with angular speed ω.

However, a problem arises when the length of wire carrying I_1 rotates into the position currently occupied by I_2. If the current is still flowing the same direction, but the positions are reversed, the rotor will actually experience a backward (clockwise) torque. This is because unless we somehow change the direction of the current, I_1 will always be pulled out of the page to the left, but in order to continue the rotation we've started, when I_1 rotates around to the position of I_2 it now needs to be pushed into the page to the right.

In general, in order to keep the rotor turning continuously, anytime the wire is on bottom half of the diagram, we want it to experience a force into the page to the right, and anytime the wire is on the top half of the diagram, we want it to experience a force out of the page to the left. To make this happen, we need to be able to change the direction of the current flowing through the wire so that when the wire is on the top of half of the diagram, the current is flowing in the direction of I_1, and when it's on the bottom of the diagram, it's flowing in the direction of I_2.

In a brushed DC motor, this is made possible with the use of a commutator, which in our simple motor is represented by a split ring that rotates with the rotor and two stationary brushes that the split ring slides against as it rotates. The gray wires feeding the brushes always carry current in a constant direction, but the split ring and brushes allow the current in the rotor loop to switch directions such that the current is always flowing in the direction of I_1 when it's on top, and in the direction of I_2 when it's on the bottom.

Based on the Lorentz Force Law, we can write the following expression for the force exerted on either I_1 or I_2:

$$\overline{F} = \overline{I}\ell \times \overline{B} \tag{4.12}$$

where $I = I_1 = I_2$ is the current flowing in the loop of wire, ℓ is the length of wire at the top or bottom of the loop (i.e., the length of the I_1 arrow in Fig. 4.10), and B is the magnetic field.

Thanks to the commutator, we know that the rotor will continue to turn in the same direction, with all of the forces acting together to rotate it counterclockwise continuously. Therefore, we can set aside the direction of the forces, and focus on their magnitude.

In our simple motor described above, the magnitude of the forces will be greatest when I_1 is directly at the top and I_2 is directly at the bottom because that's when they feel the full benefit of the forces into and out of the page. When the rotor is perpendicular to this position, that is, when the brushes reach the split in the split ring, we actually want to be pushing I_1 down and I_2 up. In this simple two-pole motor, we're unable to generate forces in this direction, so we simply let the rotor coast through this part of its rotation.

In any case, at the moment of maximum force, the force described by Eq. (4.12) is acting on two wires working in concert with each other, so we can multiply by two to find the total force exerted on the pair:

$$F = 2I\ell B \tag{4.13}$$

In actuality, the rotor will be made up of many windings, so the force will be further multiplied by the number of windings, n:

$$F = 2nI\ell B \tag{4.14}$$

Torque is equal to force times the radius at which a force is acting, so at the moment of maximum force, we can write the following expression for the torque:

$$T = 2nI\ell rB \tag{4.15}$$

Interestingly, $2\ell rB$ has another important meaning in this context. In our simple motor with rectangular windings, $2\ell r$ is the area inside the windings (ℓ is one side of the rectangle, and r is one half of the other side). Given that magnetic flux (Φ) is defined as area times magnetic field, $2\ell rB$ is equal to the magnetic flux through the windings when they are perpendicular to the electric field (at the moment of the rotor is coasting through the split in the rings).

Therefore, we can rewrite Eq. (4.15) as follows:

$$T = nI\Phi \qquad (4.16)$$

In practice, we'll replace n with the so-called motor constant, K_m, which not only accounts for the number of windings but also addresses a variety of other imperfections in our analysis.

$$T = K_mI\Phi \qquad (4.17)$$

It's important to note that Eq. (4.17) is only a first approximation for the torque generated by the motor. In Sec. 4.8, we will study Faraday's Law and discover that it gives rise to a force called the back electromotive force (back EMF) that reduces the available torque compared to Eq. (4.17).

Similarly, the simple motor described in this section is only an approximation of a real-world DC motor. Its design can be used to guide the construction of an experimental motor (and in turn, a model electric car), as described in App. A. But DC motors in the marketplace are significantly more complex than this simple design, with additional sets of magnetic poles, commutators, or windings. Nevertheless, regardless of how complicated they might look in comparison to our simplified model, brushed DC motors all operate on the same basic principles that we have explored in this section.

4.6   Brushless Motors

While brushed DC motors are relatively easy to understand, in practice, it's difficult to design high-reliability long-life commutators that can withstand the abuse of daily, high-power operation in an electric vehicle.

Brushless motors solve this problem by elegantly eliminating the need for a mechanical commutator. Whereas in a brushed motor, we make the rotor turn by changing the direction of the current in the rotor windings, in a brushless motor, we make the rotor turn by rotating the magnetic field around the rotor. Given that the rotor turns in synchrony with the rotating magnetic field, brushless motors are also known as synchronous motors. In addition, when the rotor is made using permanent magnets, they are also known as permanent magnet motors. In this section, we'll explore two different kinds of brushless motors: brushless DC motors and brushless AC motors.

4.6.1   Brushless DC Motors

Figure 4.11 illustrates the operation of a brushless DC motor in two panels. In each panel, we see two sets of Helmholtz coils set

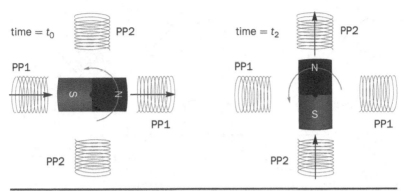

FIGURE 4.11 Operation of a brushless DC motor.

perpendicular to each other, which we'll refer to as pole pair 1 (PP1) and pole pair 2 (PP2). Inside these pole pairs, there is a permanent magnet rotor, which has a permanent north pole and south pole that are continuously trying to align themselves with the magnetic field.

The basic idea of a brushless DC motor is that we will turn the current in the pole pairs on and off such that the north pole of the stator shifts from the right coil, to the top coil, to the left coil, to the bottom coil, causing the rotor to turn counterclockwise in synchrony with the rotating magnetic field.

In the left panel, at time t_0, current is flowing through PP1 in a counterclockwise fashion (assuming the clock face is pointing to the right). Based on the curled-finger version of the right-hand rule, we can see that the magnetic field in the stator points to the right. Therefore, the north and south poles of the rotor align themselves to it.

In the right panel, at time t_2, current is no longer flowing through PP1, so there is no longer a horizontal magnetic field. Instead, current is now flowing through PP2 in a counterclockwise fashion (assuming the clock is on the floor facing up). Based on the right-hand rule, the magnetic field now points up, and the rotor once again aligns itself to this magnetic field.

To keep the rotor turning from there, we run current through PP1 in a *clockwise* fashion, causing the magnetic field to point to the left, and then through PP2 in a *clockwise* fashion, causing it to point down. Then we repeat from the beginning, causing it to rotate from the right to the top to the left to the bottom, always pulling the rotor around with it.

Figure 4.12 shows the current applied to PP1 and PP2 over time. The square waves in black represent the magnitude and sign of the current applied to each pole respectively. The gray lines show that the current applied to PP1 is a very rough approximation of a cosine wave, and the current applied to PP2 is a very rough approximation of a sine wave ($\cos \omega t$ and $\sin \omega t$ to be exact, where ω is the angular speed, t is time, and therefore, ωt is the current angle of the rotor). As

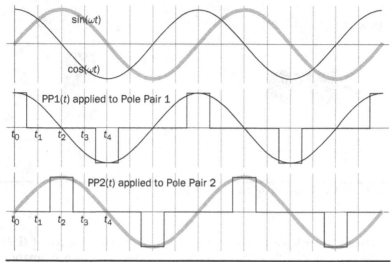

FIGURE 4.12 Current applied to a brushless DC motor.

is the nature of cosine and sine waves, they are shifted 90° (or $\pi/2$ rad) from each other, which is what allows our four-pole motor, with poles spaced 90° from each other, to function.

At time t_0, we apply a positive current to PP1, causing the magnetic field to point to the right. At time t_1, no current is applied to either pole pair, and, assuming the rotor is already rotating, the north pole of the rotor coasts through the upper right quadrant. At time t_2, we apply a positive current to PP2, causing the magnetic field to point up. (If the motor wasn't already turning at time t_0, the rotor would start moving now to align itself with PP2.) At time t_3, no current is applied, and the north pole of the rotor coasts through the upper left quadrant. At time t_4, we apply a negative current to PP1, causing the magnetic field to point to the left. After that, the rotor coasts through the bottom right quadrant, then we apply a negative current to PP2 to turn it another 90°, then it coasts through the upper right quadrant, and finally, we repeat the whole cycle from the beginning.

Of course, the motor described above only works if the pulses of current through PP1 and PP2 are fired at exactly the right time. Therefore, to keep the pulses synchronized with the rotation of the rotor, brushless DC motors contain sensors that detect the present angle of the rotor and allow the controller (which we'll learn about in Chap. 6) to fire the pulses at the proper time to generate smooth and continuous rotation.

4.6.2 Brushless AC Motors

In the case of a brushless DC motor, we're using DC current to roughly approximate AC current, or alternating current, which varies over time

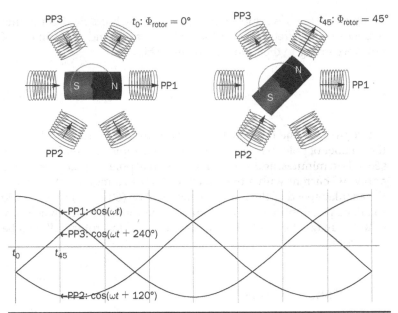

FIGURE 4.13 Operation of a brushless AC motor.

as defined by sine and cosine. So what if, instead of using DC current, we simply use AC current?

Consider the brushless AC motor shown in Fig. 4.13. Now, rather than two pole pairs, it has three pole pairs. And rather than two different approximations of AC current, we are applying three phases of actual AC current, each shifted by 120° relative to the next. The first phase of AC power is applied to PP1, the second (+120°) to PP2, and the third (+240°) to PP3.

At time t_0, the current applied to PP1 is at a maximum, so the rotor is aligned with the strongest magnetic field, which is pointing to the right. As time progresses, the field generated by PP1 decreases, and the field generated by PP2 increases. (The current to PP2 is negative at this time, but it's heading toward its largest negative value, so the field strength toward the upper right is increasing.) The upper right panel of Fig. 4.13 shows a point in the middle of this process, t_{45}, where the rotor has rotated 45°, and is heading toward PP2 as it approaches its maximum field strength, which will occur when the angle of the rotor is 60°.

Similar to the brushless DC motor, the shifting currents to each of the pole pairs cause the strongest point of the magnetic field to rotate counterclockwise, pulling the permanent magnet rotor along with it. By applying continuous AC current to three pole pairs, we can generate an even smoother rotation of the magnetic field than we can with our previous two DC pole pairs which were only crudely approximating AC current.

In a brushless AC motor, the speed at which the field is rotating depends on the frequency of the AC waveform and the number of poles according to the following relationship:

$$f_{\text{synch, RPM}} = \frac{60 \times 2 \times f_{\text{AC, Hz}}}{N_{\text{poles}}} \tag{4.18}$$

where $f_{\text{AC, Hz}}$ is the frequency of the waveform in hertz, and N_{poles} is the number of poles (six, in the case of our example above). 60 converts seconds to minutes, and 2 converts individual poles to pole pairs. If we apply AC current with a frequency of 60 Hz to a motor with 6 poles, the angular speed of the magnetic field will be $(60 \times 2 \times 60)/6 = 1{,}200$ rotations per minute (rpm). Given that the rotor turns in synchrony with the magnetic field, the angular speed of the rotor will also be 1,200 rpm.

4.7   Reluctance Motors

While most synchronous motors use a permanent magnet rotor, there's another intriguing option that Tesla has started to use in some of its vehicles: the reluctance motor.

Consider what happens when you place a magnet near a piece of magnetic material: not something that has already been magnetized, just something that magnets are able to stick to, like iron. Why do magnets stick to this material? When a magnet gets close to it, the magnetic poles of that material align themselves to the magnetic field of the magnet, creating a mutual attraction. This is the fundamental principle underlying reluctance motors.

Consider the rotor and stator illustrated in Fig. 4.14. Starting in the position illustrated on the left side of the figure, we energize PP1, creating a magnetic field between the poles at the bottom right and

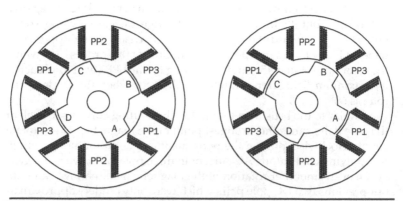

Figure 4.14   Operation of a reluctance motor.

top left. These energized stator poles attract the closest poles of the iron core, in this case, A and C, ending up in the position illustrated on the right. From here, we energize PP2, which attracts the nearest poles of the iron core, in this case, B and D. From there, we energize PP3, attracting poles A and C, then PP1, attracting poles B and D, etc.

Note that this reluctance motor only works because the number of stator poles and rotor poles is mismatched. In this case, there are six stator poles and four rotor poles. This ensures that at every moment, there are rotor poles and stator poles out of alignment, allowing them to be attracted to each other to create rotation. If the numbers of poles were equal, and the poles were aligned, firing any of the stator poles would simply pull outward on the rotor poles directly in front of them, which isn't useful for generating rotation. Even though the number of stator and rotor poles is mismatched, the rotor still turns at the same speed as the rotating magnetic field, so reluctance motors are also considered synchronous motors.

While reluctance motors rely primarily on electromagnetic stator poles and a rotor that is magnetic, but not a permanent magnet, the reluctance motors that Tesla is using are said to be permanent magnet reluctance motors. This potentially confusing name simply refers to the fact that they use a small amount of permanent magnet in the stator poles to smooth out the rotation of the motor, which can otherwise be choppier than we might like for use in an electric vehicle.

Most electric vehicles on the market today use some kind of synchronous motor, with a rotating magnetic field in the stator that pulls a magnetic rotor around with it. But there's another important type of motor in use today that uses a similar stator, but a different kind of rotor, one which doesn't use magnetic material. Instead, these motors work by inducing a current in the rotor that responds to the rotating magnetic field created by AC current applied to the stator. For this reason, they're called AC induction motors. To appreciate how AC induction motors work, we must first study Faraday's Law.

4.8   Faraday's Law

So far, we've seen that a current traveling through a uniform magnetic field experiences a force (Lorentz Force Law), and a current traveling through a wire generates a magnetic field (Biot-Savart's Law). Based on these principles, we have shown several different ways that electromagnetic forces can be used to generate rotational motion.

In this section, we will study Faraday's Law, which states that a change in the magnetic flux flowing through a loop of wire induces a voltage in that loop. This will have several important implications. First, it means that in addition to turning electricity into rotational motion, motors can also be used as generators to turn rotational motion into electricity. This is what enables regenerative braking in electric vehicles. Second, it gives rise to the back EMF, foreshadowed in

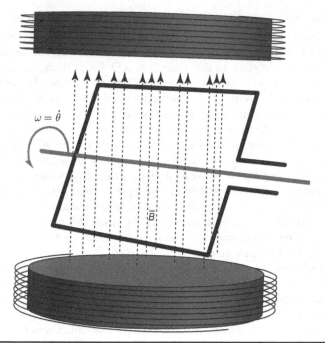

FIGURE **4.15** A single winding inside an electric generator.

Sec. 4.5, which will reduce the torque available from the motor as compared to the first approximation we developed in that section. Finally, it is what enables AC induction motors, which we'll introduce in Sec. 4.9.

Faraday's Law can be expressed as follows:

$$V = -\frac{d\Phi}{dt} \tag{4.19}$$

where V is the voltage induced in the loop of wire, and Φ is the magnetic flux that flows through the surface area defined by the loop of wire. The derivative, $d\Phi/dt$ measures the change in that magnetic flux over time. Figure 4.15 shows a single winding inside an electric generator. If this looks familiar, it's because it is: this setup of a Helmholtz coil surrounding a loop of wire was exactly what we used to construct our simple brushed DC motor back in Sec. 4.5.

The Helmholtz coil is generating a uniform vertical magnetic field, $\overline{B}$, shown by the dashed lines. The loop of wire spins counterclockwise in the magnetic field due to torque applied from the outside, for example, torque from the rotating wheels of an electric vehicle traveling down a hill. Of course, in the real world, this loop of wire would actually be many windings, but a single loop will suffice to develop our understanding.

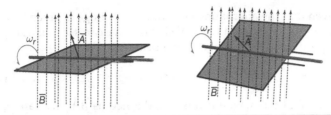

Figure 4.16 A single winding in an electric generator, angled at 20° and 50°.

Given that magnetic flux is magnetic field times the area it's passing through, we can rewrite Faraday's Law as:

$$V = \frac{-d(\overline{B} \bullet \overline{A}_r)}{dt} \qquad (4.20)$$

where $\overline{B}$ is the magnetic field, and $\overline{A}_r$ is the area enclosed by the loop of wire. Given that the magnetic field between the two poles of the Helmholtz coil is essentially uniform, we will treat B as constant, and focus on the change in A_r over time.

Figure 4.16 shows the loop of wire at two different angles, at 20° from perpendicular to the field, and at 50° from perpendicular to the magnetic field. At 20°, the area enclosing the magnetic field is greater than the area enclosing the magnetic field at 50°. To see why, imagine holding a sheet of paper under a bright light that's directly overhead. If you hold the paper parallel to the floor, it casts a large shadow. But as you rotate the paper, tilting it so the right side is higher than the left, the shadow it casts on the floor will get smaller, until, when the paper is directly perpendicular to the ground, its shadow will essentially disappear. The area of the shadow on the floor is the area we are interested in when it comes to this analysis.

Because the area that the magnetic field is passing through varies as a function of time from a maximum at 0° to 0 at 90°, we can express the time-varying value of this area as A_r when it's at its maximum times $\cos(\omega_r t)$. Cosine is the proper function to use because cosine is also at its maximum at 0° and 0 at 90°. Therefore, we can write Faraday's Law as follows:

$$V = \frac{-d[BA_r \cos(\omega_r t)]}{dt} \qquad (4.21)$$

Given that the derivative of $\cos(\omega_r t)$ is $-\omega_r \sin(\omega_r t)$, we can rewrite this as:

$$V = BA_r \omega_r \sin(\omega_r t) \qquad (4.22)$$

If the rotor is made up of not just one loop, but n windings, the induced voltage is multiplied by n, as follows:

$$V = nBA_r \omega_r \sin(\omega_r t) \qquad (4.23)$$

Given that the alternating current (AC) voltage is given by a sine wave, and the maximum value of sine is 1, we can see that our generator is producing AC voltage with a peak value of:

$$V = nBA_r\omega_r \tag{4.24}$$

Given that the magnetic flux Φ is defined as BA_r, we can replace BA_r with Φ:

$$V = n\Phi\omega_r \tag{4.25}$$

Finally, as we did in Eq. (4.17), in the real world, we replace n with the motor constant, K_m, which accounts not only for the number of windings, but also addresses other a variety of other imperfections in our analysis:

$$V = K_m\Phi\omega_r \tag{4.26}$$

There are two important implications to this analysis. First, we have seen that torque applied from the outside, in this case, torque produced by vehicle motion, generates a voltage that can be used to recharge the batteries of the vehicle.

Second, this also has implications for the torque available under normal vehicle operation, when the motor is converting electrical energy to rotational energy. In this case, the voltage defined by Eq. (4.26) opposes the voltage applied by the battery. For this reason, this voltage is called the back EMF, or V_B. Given that the back EMF voltage increases as a function of rotational speed, ω_r, the torque available from the motor will decrease as a function of ω_r.

We can quantify this effect by recalling Eq. (4.17) from our analysis of brushed DC motors:

$$T = K_mI\Phi \tag{4.27}$$

The current, I, flows in the rotor of the brushed DC motor. Based on Ohm's Law, $V = IR$, this current is equal to the applied voltage divided by the resistance of the rotor. In this case, the applied voltage is the source voltage from the battery, V_S, minus the back EMF voltage, V_B:

$$I = \frac{V_S - V_B}{R_r} \tag{4.28}$$

If we put in the value of V_B from Eq. (4.26), we get:

$$I = \frac{V_s}{R_r} - \frac{K_m\Phi\omega}{R_r} \tag{4.29}$$

Substituting this whole expression for I from Eq. (4.29) for I in Eq. (4.27), we get:

$$T = \frac{K_m\Phi V_s}{R_r} - \frac{(K_m\Phi)^2\omega}{R_r} \tag{4.30}$$

This important relationship is plotted for a 10-kW DC motor in Fig. 4.17. The diagonal line shows the relationship defined by Eq. (4.30).

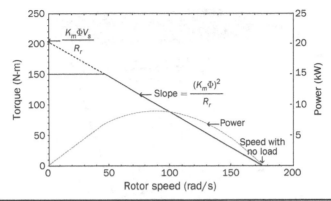

Figure 4.17 Torque and power as a function of speed for a 10-kW DC motor.

In the upper right, you can see that theoretically, a maximum of 200 N-m of torque is available at a rotor speed of 0, but in practice, this low-speed torque is limited to 150 N-m by the controller to protect the motor from high currents and mechanical fatigue due to high torque. As the rotor speed increases, the back EMF increases, limiting the available torque. Eventually, the motor reaches a speed where the available torque is zero.

The dashed curve at the bottom of Fig. 4.17 shows the mechanical power available from the motor as defined by the fundamental equation for power in rotational systems:

$$P = T\omega \tag{4.31}$$

in which P is power, T is torque, and ω is rotational speed. In the case of our DC motor above, at low speed, power increases as the speed increases, but it eventually decreases to zero at high speed as the back EMF reduces the available torque.

4.9 AC Induction Motors

In Sec. 4.6.2, we studied brushless AC motors, which work by using three-phase AC power to generate a continuously rotating magnetic field that pulls the magnetic poles of the rotor around in perfect synchrony with the rotating field.

The stator of an AC induction motor works exactly the same way, generating a continuously rotating magnetic field. But the rotor of an induction motor is a bit more interesting. Instead of being magnetically charged, the rotor of an AC induction motor is simply a collection of loops of wire.

In Sec. 4.8, we saw that when a loop of wire rotates through a uniform magnetic field, a current is induced in that loop of wire. But the opposite is also true: when a magnetic field rotates around a loop of wire, a current is also induced. Then, as a result of the Lorentz Force

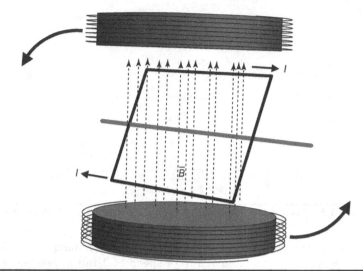

Figure 4.18 Understanding an induction motor by analogy to a generator.

Law, the loop of wire experiences a force. In the case of an AC induction motor, this force is a torque in the direction of the rotation of the magnetic field.

Why is this? Consider once again the single set of magnetic poles and loop of wire in Fig. 4.15, reproduced with new labels in Fig. 4.18. Imagine that, instead of the loop of wire rotating counterclockwise in between stationary poles, the magnetic poles are rotating counterclockwise around a stationary loop of wire. It's important to note that in the real world, the physical coils of the stator will be stationary, it's only the magnetic field that's rotating, but by way of analogy to Fig. 4.15, it will be easiest to visualize if we imagine that the poles in that diagram are rotating. *Relative to the counterclockwise motion of the poles*, the stationary loop of wire is now rotating clockwise.

As we saw in Sec. 4.8, Faraday's Law states that the electromotive force works in opposition to the rotational motion of the loop of wire. Given that the stationary loop is moving clockwise relative to the magnetic field, the electromotive force works to turn the loop counterclockwise to oppose this relative motion.

In this particular example, it does so by inducing a current in the loop that travels into the page through the bottom wire and out of the page through the top wire. Using the right-angle right-hand rule, you can see that the $\bar{I} \times \bar{B}$ pulls the top wire out of the page to the left and pushes the bottom wire into the page to the right, turning the loop counterclockwise.

Unlike the other motors we've studied so far, the rotation of the rotor in an AC induction motor is not in synchrony with the rotation of the magnetic field. In fact, in order for the motor to work, it can't be. The only reason that induction motors work is that there is relative motion between the rotation of the loops of wire in the rotor and the

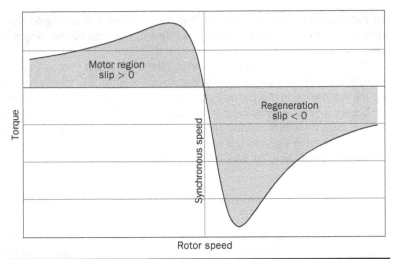

FIGURE 4.19 Torque as a function of rotor speed for an AC induction motor.

rotation of the magnetic field of the stator. If the rotor turned in perfect synchrony with the magnetic field, there would be no change in magnetic flux through the loops of wire, which means no induced current, and no torque. Therefore, in order for an induction motor to function, the rotational speed of the rotor must be slower than the rotational speed of the magnetic field to maintain the relative motion required for induction. In the context of an induction motor, this lag between the rotation of the rotor and the magnetic field is known as *slip*.

The slip of the rotor, s, is defined as a percentage:

$$s = 100 \left(1 - \frac{f_{\text{Rotor}}}{f_{\text{Synch}}} \right) \tag{4.32}$$

where f_{Rotor} is the rotational speed of the rotor and f_{Synch} is the rotational speed of the magnetic field in the stator, also known as the synchronous speed. Typically, induction motors operate with a few percent of slip.

Figure 4.19 shows the relationship between slip and torque for a sample induction motor. The left half of the figure illustrates the torque available when the rotor speed is slower than the synchronous speed. As we previously noted, the torque is zero when the rotor speed matches the synchronous speed, but rises to a maximum when the rotor speed is slightly less than the synchronous speed.

The right half of the figure illustrates what happens when the rotor speed is faster than the synchronous speed. This is only possible when the motor is experiencing an externally applied torque, like that provided by the wheels of an electric vehicle traveling down a hill. In this situation, the AC induction motor is actually acting as an AC induction generator and converting this external torque into electric power to recharge the batteries. Given that the generator is acting to slow down the rotor, which is getting ahead of it, the sign of the torque is negative.

Of course, in the operation of an electric vehicle, we will need to reach a wide variety of operating points within the speed/torque space that aren't on this particular curve: maybe we need higher speed, or higher torque. Fortunately, if we recall Eq. (4.18) from Sec. 4.6.2, we can change the synchronous speed by changing the frequency of the AC waveform applied to the stator.

For convenience, Eq. (4.18) is reproduced below:

$$f_{\text{synch, RPM}} = \frac{60 \times 2 \times f_{\text{AC, Hz}}}{N_{\text{poles}}} \tag{4.33}$$

By changing the frequency of the AC waveform, we can change the synchronous speed, and therefore the rotor speed and available torque. Let's take a look at a particular example. Figure 4.20 shows the various torque vs. speed curves that are available for a variable-speed 18-kW AC induction motor. Each curve corresponds to a particular frequency of AC waveform that is being applied to the stator, in 10-Hz increments from 10 Hz to 70 Hz. Note that each curve includes only the left half of the curve from Fig. 4.19, where the motor is acting as a motor, not a generator.

As you can see, by changing the AC waveform, we can reach a greater number of operating points in the speed/torque space. By modulating the waveform even more precisely, say, in 5-Hz increments, or 1-Hz increments, we could reach an even greater number of operating points.

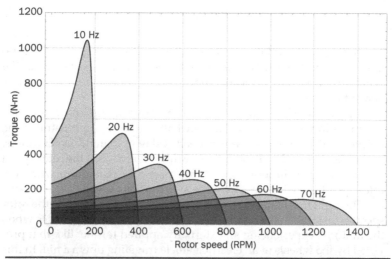

FIGURE 4.20 Family of torque vs. speed curves for a small AC induction motor.

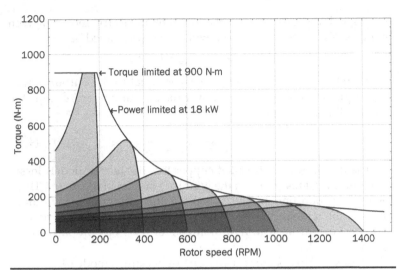

FIGURE 4.21 Family of torque vs. speed curves with limits.

Figure 4.21 shows the same curves as Fig. 4.20, but with two important limits added. At low rotor speed, the peak of the first curve is limited by the 900 N-m torque limit of this particular motor, and at high rotor speed, the peaks of the curves are limited by the 18-kW power limit of the motor.

In the end, the total operating space of the motor as defined by these individual curves (and others like them) is very similar to the torque vs. speed curve of the DC motor in Fig. 4.17. In both cases, there is high (and externally limited) torque available at low speed, but this torque decreases as the rotor speed increases.

4.10   Motor Efficiency

An important consideration for any type of motor is its efficiency, η_m, which is equal to the output power we get out of the motor divided by the input power we put into it. Unfortunately, this value is never 100 percent.

$$\eta_m = \frac{\text{output power}}{\text{input power}} \qquad (4.34)$$

The input power is equal to the output power plus a variety of different kinds of losses that we'll explore below.

$$\eta_m = \frac{\text{output power}}{\text{output power} + \text{losses}} \qquad (4.35)$$

A good approximation of the losses in an electric motor is given by Eq. (4.36), the terms of which will each be explained below:

$$\text{Losses} = k_c T^2 + k_i \omega + k_w \omega^3 + C \tag{4.36}$$

Therefore, the motor efficiency η_m is equal to the output power, $T\omega$, torque times angular speed, over the output power, $T\omega$, plus the losses defined in Eq. (4.36).[1]

$$\eta_m = \frac{T\omega}{T\omega + k_c T^2 + k_i \omega + k_w \omega^3 + C} \tag{4.37}$$

The first loss, $k_c T^2$, is called copper loss because it models losses due to electrical resistance in the windings of wire in the motor. These losses are proportional to $I^2 R$ where I is the current in the windings and R is the resistance of the windings. Given that torque is proportional to current, we can represent copper losses as proportional to torque squared.

The second loss, $k_i \omega$, is called iron loss because it models losses due to the iron components in the motor. There are two main causes of iron losses: (1) it takes energy to continually magnetize and demagnetize the iron components of the motor as the magnetic field changes, and (2) this changing magnetic field induces a current in the iron, which heats the iron. Both effects increase as the frequency of the magnetic field changes increases, so the iron loss is proportional to the angular speed, ω.

The third loss, $k_w \omega^2$, is called windage loss, and represents aerodynamic resistance on the rotor. As we learned in Chap. 3, the power required to overcome aerodynamic resistance is proportional to speed cubed, so the windage loss is proportional to ω^3.

The fourth loss, C, includes constant losses that do not vary with speed or torque, such as the power required to maintain the magnetic field when the motor is stationary and the power required to run the controller.

If we know the values of k_c, k_i, k_w, and C, we can plot η_m as a function of T and ω. To this end, Fig. 4.22 shows the efficiency curves for

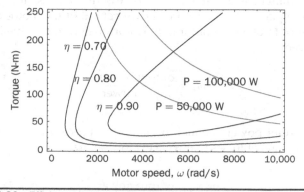

FIGURE 4.22   Efficiency curves for a 100-kW AC induction motor.

a 100-kW AC induction motor with $k_c = 0.3$, $k_i = 0.01$, $k_w = 5 \times 10^{-5}$, and $C = 600$. It also includes curves that show where the motor is operating in this efficiency space when it's running with 50 kW or 100 kW of power.

The induction motors in the Tesla Model S and Model X have a reported efficiency of 93 percent, while the permanent magnet reluctance motors in the Model 3 and Y have a reported efficiency of 97 percent. In the near future, Tesla plans to upgrade the S and X to run on these higher efficiency reluctance motors as well.[2]

4.11 Maximum Acceleration

One of the most desirable features of an electric car is how quickly and smoothly it can accelerate in response to the torque from its electric motor. In fact, exactly how fast a given electric car can accelerate is largely defined by the power of its motor.

If one knows the intimate details of a car's drivetrain, it's possible to calculate its maximum acceleration from first principles, as Larminie and Lowry (2012) do. However, as Masias (2018) notes, it's not actually necessary to go through the trouble of doing this—which is a good thing, because the drivetrain parameters required for Larminie and Lowry's style of calculation are difficult to come by.

As Masias (2018) astutely observes, the time it takes to accelerate a car from 0 to 60 mph can be approximated as a linear function of the maximum power output of its motor(s) (in kW), and the weight of the car (in kg).[3] We can clearly see this in Fig. 4.23, which plots the 0 to 60 time of every major EV currently for sale in the United States as a function of its weight to power ratio (in kg/kW), from the Tesla Model S in the lower left to the Kia Soul EV in the upper right.

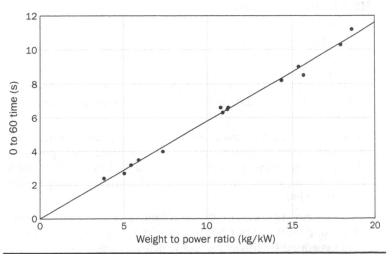

FIGURE 4.23 0 to 60 acceleration time as a function of weight-to-power ratio.

Conveniently, all of the available data is clustered tightly around a linear trendline with a slope of 0.58 and an intercept that is indistinguishable from 0. This means that in order to find the 0 to 60 time of a newly designed EV, all we need to do is enter its weight and the maximum power output of its motor(s) into the following formula:

$$T = 0.58 \frac{W}{P} \qquad\qquad (4.38)$$

where T is the 0 to 60 time in seconds, W is the weight of the car in kg, and P is the power of its motor (or the combined power of its motors) in kW.

In addition to allowing us to approximate the 0 to 60 time of any electric car, Eq. (4.38) also gives us guidance on designing new electric cars: if we want our car to accelerate faster, we need to either decrease the weight of the car, increase the power of the motor, or both.

4.12   Conclusion

This chapter has covered a huge amount of ground, building our understanding of motors from basic principles in physics (the Lorentz Force Law, Coulomb's Law, Biot-Savart's Law, and Faraday's Law) to the engineering and operation of specific types of motors (Brushed Motors, Brushless Motors, Reluctance Motors, and Induction Motors). In addition, we have seen how motors can be used as generators, enabling regenerative braking, and how back EMF limits torque at high speeds. We also developed an expression for motor efficiency and a formula for calculating maximum acceleration.

4.13   Homework Problems

4.1 Imagine that you are standing directly on top of the magnetic North Pole of the Earth, where the magnetic field lines point straight down into the ground (the magnetic North Pole actually corresponds to the south pole of a magnet). If you fire a charged particle straight forward away from you into this magnetic field, what path will the particle take? (*Hint*: Compare this scenario to Fig. 4.2.)

4.2 Imagine that you are now standing on the magnetic South Pole of the Earth, where the magnetic field lines point straight up into the sky. If you are holding a wire carrying current from your right to your left, which way will the wire be pulled? (*Hint*: Compare this scenario to Fig. 4.3.)

4.3 Imagine once again that you are holding a wire carrying current from your right to your left. What is the shape of the magnetic field generated by this wire, and which way does it point? (*Hint*: Compare this scenario to Fig. 4.5.)

4.4 Imagine that you're looking straight down the center of a coil of wire that is carrying current counterclockwise. What is the shape of the magnetic field generated by this wire, and which way does it point? (*Hint*: Compare this scenario to Fig. 4.6.)

4.5 Consider the brushed DC motor illustrated in Fig. 4.24. Which way is the rotor turning? (*Hint*: Compare this scenario to Fig. 4.10.)

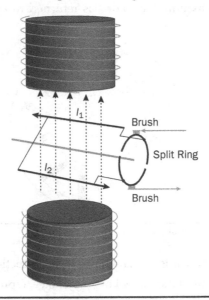

FIGURE 4.24   Analysis of a brushed DC motor.

4.6 Consider the brushless DC motor illustrated in Fig. 4.25. Given that $\cos \omega t$ is applied to PP1 and $\sin \omega t$ is applied to PP2, what direction will the north pole of the rotor point at the time indicated by the bold vertical line?

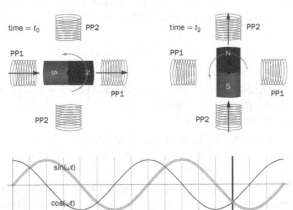

FIGURE 4.25   Analysis of a brushless DC motor.

4.7 A brushless AC motor with 6 pole pairs (12 poles) is operating at an AC frequency of 50 Hz. What is the synchronous speed of the motor in RPM?

4.8 Consider the reluctance motor illustrated in Fig. 4.26. The rotor is currently at rest. What will happen if we energize PP1? What if we energize PP2? What about PP3? (Between each of these three questions, assume the rotor has returned to rest in its original position.)

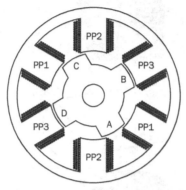

FIGURE 4.26   Analysis of a reluctance motor.

4.9 In your own words, explain how an AC induction motor works.

4.10 Using a computational tool of your choice, reproduce Fig. 4.22.

4.11 Unofficial estimates put the weight and power of Tesla's Tri-Motor Cybertruck at about 3,000 kg and 600 kW, figures which are perfectly in line with its reported 0 to 60 time of 2.9 sec based on Eq. (4.38). Its towing capacity is reported to be at least 14,000 lb. Assume a Tri-Motor Cybertruck is towing three Ford F-150s weighing approximately 4,500 lb each. Based on Eq. (4.38), how long will it take for the Cybertruck and its train of three F-150s to accelerate from 0 to 60 mph? (*Hint*: Be careful with units.)

Notes

1. This analysis of motor efficiency is based on the one presented in: Larminie, James and John Lowry. *Electric Vehicle Technology Explained*. Hoboken, NJ: Wiley, 2012.

2. Lambert, Fred. "Tesla is Upgrading Model S/X with New, More Efficient Electric Motors." *Electrek*, April 5, 2019. https://electrek.co/2019/04/05/tesla-model-s-new-electric-motors/

3. Masias, Alvaro. "Lithium-Ion Battery Design for Transportation." In *Behaviour of Lithium-Ion Batteries in Electric Vehicles: Battery Health, Performance, Safety, and Cost*, edited by Gianfranco Pistoia and Boryann Liaw, 1–34, Cham, Switzerland: Springer, 2018.

General references for this chapter include:

Fitzgerald, A. E., C. Kingsley, and A. Kusko. *Electric Machinery*, 3rd edition. New York: McGraw-Hill, 1971.

Haus, H. A., and J. R. Melcher. *Electromagnetic Fields and Energy*. Englewood Cliffs, NJ: Prentice Hall, 1989.

Larminie, James and John Lowry. *Electric Vehicle Technology Explained.* Hoboken, NJ: Wiley, 2012.

Image Credits

CHAPTER 5
Batteries

5.1 Introduction

In order for our electric vehicle motor to run, we need a portable form of electricity. Most often, this electricity is stored in a battery. In this chapter, we will take a look at the key components and characteristics of electric vehicle batteries, also known as traction batteries. Because nearly every electric vehicle on the market today uses lithium-ion batteries, we will focus on this particular chemistry.

5.2 Battery Fundamentals

First used in this way by Benjamin Franklin in 1749, the word "battery" refers to a group of electrochemical cells. An electrochemical cell is a device that stores energy in the form of chemical potential energy, then releases it as electricity.

The smallest electrochemical unit that is capable of providing power is called a cell. As the voltage of a single cell—generally 3.6 V for lithium-ion batteries—is nowhere near sufficient to power a full-size electric vehicle, individual cells are combined to form modules, and individual modules are combined to form battery packs.

Each cell is made up of a positive electrode, called the cathode, and a negative electrode, called the anode. Between the cathode and anode, there is an electrolyte that allows positive ions to travel back and forth between the two electrodes, and a separator that prevents the electrodes from coming in contact with each other and short circuiting. When connected to a circuit, electrons travel back and forth between the cathode and the anode through the circuit.

Early electric vehicles ran on primary cells, which means they were not rechargeable. The battery began with a fixed amount of chemical energy, and once it was all converted to electricity, that was it. To get the battery working again, you needed to replace the reactants (similar to refueling an ICE vehicle with gasoline) or swap out the battery for an entirely new one. Later, these primary batteries were replaced with rechargeable batteries, also called secondary batteries or storage batteries.

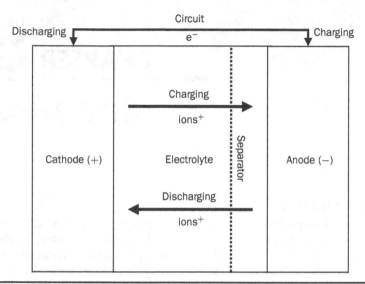

FIGURE 5.1   Basic components of a battery.

When a storage battery is charging, electrons travel from the cathode to the anode through the circuit while positive ions travel in parallel through the electrolyte. When the battery is discharging, the process is reversed, as electrons and positive ions travel through the circuit and electrolyte from the anode to the cathode (Fig. 5.1).

5.3   Lithium-Ion Batteries

Since the vast majority of electric vehicles on the market today use lithium-ion batteries, we will focus on this particular chemistry. We will first take a look at the components that make up the battery, followed by the processes that make it run.

5.3.1   Cathode ($+$)

The cathode of a lithium-ion battery is made of a lithium-containing compound: most commonly, lithium cobalt oxide ($LiCoO_2$). Other cathode materials include lithium manganese oxide ($LiMn_2O_4$), lithium iron phosphate ($LiFePO_4$), lithium nickel cobalt aluminum oxide ($LiNiCoAlO_2$), or lithium nickel manganese cobalt oxide ($LiNi_xMn_yCo_zO_2$). Given that the basic processes are essentially the same, we will use lithium cobalt oxide as our example.

The cathode is the natural home of the Li^+ ions (and corresponding electrons) that make the battery run. When the battery is fully discharged (and when the battery is being manufactured), the cathode is where the lithium lives. On charging, the Li^+ ions take a vacation to the anode. On discharging, the Li^+ ions return home to the cathode.

5.3.2 Anode (−)

The anode of a lithium-ion battery is made of a compound that can temporarily accept Li^+. Most commonly, it is made of graphite (C_6). Other anode materials include lithium titanate ($Li_4Ti_5O_{12}$), hard carbon, an alloy of tin and cobalt, or a combination of silicon and carbon. Given that the basic processes are essentially the same, we will use graphite as our example.

The anode is like a vacation home for the Li^+ ions. On charging, the anode temporarily accepts them in, only to send them home to the cathode on discharging.

5.3.3 Electrolyte

The electrolyte is what allows the Li^+ ions to move between the cathode and the anode. The key characteristic of the electrolyte is that it is an ionic conductor and an electrical insulator. This means that lithium ions can easily pass through it, but electrons can't.

In lithium-ion batteries, the electrolyte generally consists of lithium salts (such as $LiPF_6$, $LiAsF_6$, $LiClO_4$, $LiBF_4$, or $LiCF_3SO_3$) dissolved in an organic solvent (such as ethylene carbonate, dimethyl carbonate, or diethyl carbonate). The electrolyte also includes additives to prevent degradation of the electrodes, just as high quality gasoline includes additives to prevent degradation of the engine.

5.3.4 Circuit

Though technically not part of the battery itself, the circuit is an integral part of its operation. Similar to (but the exact opposite of) the electrolyte, the circuit is an electrical conductor and an ionic insulator— that is, electrons can easily pass through it, but lithium ions can't. This means that while lithium ions travel from one electrode to the other through the electrolyte, a corresponding amount of electrons travel from one electrode to the other through the circuit. It's like the electrolyte and the circuit are a boat and a train. When traveling from home (cathode) to vacation home (anode), or vice versa, the Li^+ ions always take the boat (electrolyte), while the electrons always take the train (circuit). They travel in parallel on different routes, meeting up at their destination.

5.3.5 Separator

The separator is what keeps the anode and cathode from touching and prevents a short-circuit. This separation of charge is what allows the battery to store energy. In lithium-ion batteries, the separator is generally a thin, porous material such as cloth made from fiberglass or film made from nylon, polyethylene, or polypropylene. The separator allows Li^+ ions to travel through the electrolyte while ensuring electrons travel through the circuit instead.

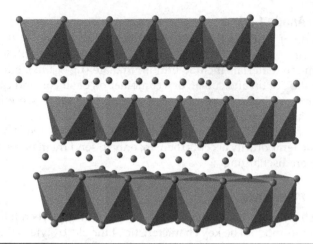

FIGURE 5.2 Lithium intercalated in CoO_2.

5.3.6 Charging

When the battery is being charged, the charger uses power from the grid to transfer electrons from the cathode to the anode. At the same time, a corresponding number of Li^+ ions move from the cathode through the electrolyte and separator to the anode.

In our example, Li^+ is extracted from lithium cobalt oxide at the cathode, and inserted into graphite at the anode. The technical terms for this extraction and insertion are deintercalation and intercalation. When we say something is intercalated, this simply means it is stored within the lattice structure of a compound. Figure 5.2 shows lithium intercalated in $LiCoO_2$. It shows three layers of CoO_2 with many individual lithium atoms intercalated between the layers. LiC_6 has a similar structure, except the C_6 layers are made of carbon rings.

Equations (5.1) and (5.2) show what is happening at the cathode and the anode during charging. At the cathode, $LiCoO_2$ gives up an Li^+ and an e^-, and at the anode, they combine with C_6 to form LiC_6.

$$\text{Cathode: } LiCoO_2 \longrightarrow CoO_2 + Li^+ + e^- \tag{5.1}$$

$$\text{Anode: } Li^+ + e^- + C_6 \longrightarrow LiC_6 \tag{5.2}$$

5.3.7 Discharging

When the battery is being used (discharged), electrons move from the anode to the cathode, powering the load. At the same time, a corresponding number of Li^+ ions move from the anode through the electrolyte and separator to the cathode.

Equations (5.3) and (5.4) show what is happening at the anode and the cathode during discharging. At the anode, LiC_6 gives up an Li^+ and an e^-, and at the cathode, they recombine to form $LiCoO_2$.

$$\text{Anode: } LiC_6 \longrightarrow Li^+ + e^- + C_6 \tag{5.3}$$

$$\text{Cathode: } CoO_2 + Li^+ + e^- \longrightarrow LiCoO_2 \qquad (5.4)$$

The chemical processes involved in other lithium-ion chemistries are similar, simply replacing CoO_2 and C_6 with the corresponding compounds that serve as the lattice structures at the cathode and anode.

Another way to think about charging and discharging is to imagine that, when charging, we're pushing the lithium ions and electrons up a hill, so to speak, to a place of higher potential energy. Discharging, in this analogy, is equivalent to letting them fall back down the hill in a controlled manner, converting their stored potential energy into usable work.

5.4 Battery Characteristics

In addition to these essential battery components and processes, every battery has several essential characteristics. In this section, we'll walk you through some of the most important of these characteristics, starting with those that are of the greatest importance to the average EV driver.

5.4.1 Cost

For the driver (or at least the owner) of an electric vehicle, one of the most important aspects of a battery is its cost, usually reported in dollars per kWh of energy storage. Although lithium-ion batteries are still quite expensive, in recent years, their cost has dropped dramatically.

According to a recent analysis by Forbes, the estimated cost of the battery in a Tesla has decreased from $230 per kWh in 2016 to $127 per kWh in 2019. This means that the cost of a 100-kWh Model S battery has fallen over $10,000 in only three years. Outside of Tesla, Forbes estimates that EV battery costs have dropped from $288 per kWh in 2016 to around $158 per kWh in 2019.[1] As battery costs continue to fall, this will allow for lower-price electric vehicles with the same range, or higher-range electric vehicles for the same price, both of which will make electric vehicle ownership attractive to an even greater number of people.

5.4.2 Energy Storage/Charge Capacity

Another essential characteristic of an electric vehicle battery is the amount of energy it can store, usually reported in kWh. All else being equal (i.e., ignoring differences in vehicle efficiency), energy storage is a rough proxy for range: in general, the more kWh stored in a car's battery, the farther that car will be able to travel. As battery costs have fallen, average energy storage has increased because automakers have been able to include significantly more energy storage for the same price.

Technically speaking, energy storage is only a rough measure. This is because while batteries appear to store energy (and in some sense they do), what they're really storing is charge (electrons and lithium ions). Therefore, in a technical context, we're often more concerned with a battery's charge capacity than its energy storage.

While the SI unit for charge is the coulomb (C), this is generally considered to be too small to be useful in real-world battery applications. Therefore, the common unit for the charge capacity of a battery is the amp-hour (Ah), which is the amount of charge carried by 1 A over the course of 1 hour (equal to 3,600 coulombs).

The charge capacity of the battery is represented by the letter C (not to be confused with coulombs). For example, the LG H2 lithium-ion cell is rated to hold 3000 mAh (or 3 Ah), which means that for this cell, C = 3 Ah.

Somewhat confusingly, C has additional meanings in the context of batteries. If someone says 3C in the context of the LG cell (with its C of 3 Ah), they mean a current of 9 A ($9 = 3 \times 3$). If they say 0.5C, they mean a current of 1.5 A ($1.5 = 0.5 \times 3$). The LG H2 is rated to provide a peak current of 11.7C (or 35 A), and a stable current of 6.7C (or 20 A).[a]

Even though charge capacity (in Ah) is preferred in technical contexts, energy storage (in kWh) is still an important measure in the real world, as it makes a huge difference to the driver whether their battery stores 100 kWh, as in the case of a Tesla Model S, or 8.8 kWh, as in the case of a Toyota Prius Prime Plug-In Hybrid.

5.4.3   Battery Life

Unfortunately, the maximum capacity of a battery degrades over time. While the battery of a new Model S can hold 100 kWh, over time, as it is charged and discharged, this maximum capacity will slowly decrease. According to the latest research, this is largely due to deterioration of the anode and cathode materials in a process similar to rusting steel.[2] Fortunately, based on a survey of real-world driver data, the batteries in a Tesla Model S have been estimated to maintain 90 percent of their capacity for over 180,000 mi.[3]

Batteries are generally considered good until they reach 80 percent of their original capacity. While this "80% rule" is widespread, it's important to note that there's no reason that an EV driver who is still satisfied with 80 percent of their original range (e.g., 80 percent of the Model 3 Long Range's 322-mi-range is 258 mi) couldn't continue to use their original battery until they reached a point where they weren't satisfied. There's nothing magical about the 80 percent number that means the battery isn't good to use anymore, as you might assume based on how some people talk about it (i.e., as a battery's "end of

---

[a]To complicate things even further, C is sometimes presented with a subscript, i.e., C_{10}. This means the current that the battery is rated to provide for a discharge time of 10 hours.

useful life"). Eighty percent isn't actually the end of a battery's useful life; it's simply a useful rule of thumb.

This is especially true given that there are many ideas about how electric vehicle batteries might be given a second life after they no longer support a satisfactory driving range. For example, retired electric vehicle batteries could be used to level out the supply and demand of electricity on the power grid, or to store solar and wind electricity to use when the wind isn't blowing and the sun isn't shining, in a macro-scale version of a Tesla Powerwall. Then, once these batteries are determined to have actually reached the end of their useful life for all purposes, there are many ideas about how best to recycle them.

5.4.4 Energy Density

There are two different kinds of energy density, both of which are of great importance in the context of electric vehicles.

Gravimetric energy density (also called specific energy) is the amount of energy that can be stored in a battery of a given mass. It has units of Wh/kg. For a given amount of energy stored (in Wh), dividing by the specific energy tells you how heavy the battery will be (in kg). For a given battery mass (in kg), multiplying by the specific energy tells you how much energy the battery can store (in Wh).

$$\text{specific energy} = \frac{\text{energy}}{\text{mass}} \rightarrow \frac{WH}{KG}$$

When thinking about specific energy, it's important to consider what mass is included in this calculation. For example, while the specific energy of the individual battery cells that Tesla uses is reported to be around 250 Wh/kg, the assembled battery pack has a much lower specific energy of around 150 Wh/kg due to the added mass of the other pack components. Nevertheless, this is a dramatic increase from around 100 Wh/kg for the original Tesla Roadster battery pack just a decade ago.

Increased specific energy means that more energy can be stored in a battery of the same weight (allowing for increased range), or the same energy can be stored in a battery of lesser weight (allowing for increased vehicle efficiency). As specific energy has increased over time, automakers have applied both of these strategies to make electric vehicles that have increased range, increased efficiency, or both.

Volumetric energy density (also called energy density) is the amount of energy that can be stored in a battery of a given volume. It has units of Wh/L. For a given amount of energy stored (in Wh), dividing by the volumetric energy density tells you how big the battery needs to be (in L). For a given amount of space inside a vehicle (in L), multiplying by the volumetric energy density tells you how much energy you can store in that space (in Wh).

$$\text{energy density} = \frac{\text{energy}}{\text{volume}} \rightarrow L$$

Similar to specific energy, it's important to consider what volume is included in this calculation. In any battery pack, a significant volume will be devoted to components that do not store energy, including the pack structure and any cooling systems.

Increased energy density means that more energy can be stored in a battery of the same size (allowing for increased range), or the same energy can be stored in a battery of smaller size (allowing for increased space for cargo or safety features). As energy density has increased, automakers have applied both of these strategies to make electric vehicles that have increased range, increased space for comfort and safety, or both.

5.4.5 Specific Power

Whereas specific energy measured the amount of energy that could be stored in a battery of a given mass, specific power measures the amount of power that can be drawn from a battery of a given mass. It has units of W/kg.

$$\text{specific power} = \frac{\text{power}}{\text{mass}}$$

Depending on the application, batteries can be designed to have a high specific energy, a high specific power, or a moderate amount of both. This is analogous to having a large water bottle with a small mouth, a small water bottle with a big mouth, or a moderately sized water bottle with a moderately sized mouth.

In EV terms, this means there is a tradeoff between the maximum power you have available for acceleration and the maximum energy you have available for range. Therefore, if you were designing an electric sports car, you might want to use batteries with slightly higher specific power (at a cost of slightly lower specific energy and shorter range), while if you were designing an efficient commuter car, you might want to use batteries with slightly higher specific energy (at a cost of slightly lower specific power and slower 0 to 60 time).

5.4.6 State of Charge and Depth of Discharge

So far, we've looked at some of the characteristics that EV drivers care about when buying a car. These next two characteristics are what EV drivers care most about when driving said car.

State of charge (SoC) is the percentage of the total battery capacity that remains after use (100% = full, 0% = empty). Thus, state of charge is analogous to the fuel gauge in an ICE vehicle.

Depth of discharge (DoD) is just the opposite: it's the percentage of the total battery capacity that has been used (0% = full, 100% = empty).

The sum of SoC and DoD equals the battery capacity:

$$\text{battery capacity} = \text{SoC} + \text{DoD}$$

Estimating the state of charge of the battery is crucial to the operation of an electric vehicle because if the estimation is wrong, the driver is liable to get stranded on the side of the road with no charge (and no easy way to recharge). Unfortunately, there is no way to directly measure the charge remaining in a battery: there is no meter that can tell you exactly how many Ah you have left. There are, however, several different ways this can be estimated:

- **Voltage Method**: The voltage of a battery decreases as its remaining capacity (i.e., SoC) decreases. Using the known relationship between voltage and SoC, you can measure the voltage and use that to estimate the SoC. As voltage also depends on current and temperature, this method can be made more accurate by correcting for these factors. Given that one of the goals of battery design is to make the voltage curve as flat as possible (to ensure consistent operation over a wide range of SoC), this makes the voltage method more difficult.

- **Current Method**: This method, also known as "coulomb counting," continuously measures the current flowing out of the battery (or into it) and integrates over time to arrive at an estimate for the cumulative charge drawn from the battery (i.e., DoD), similar to what we did in Chap. 3. Of course, this method isn't perfectly accurate and therefore suffers from drift over time. The accuracy of this method can be increased by periodically resetting the counter, for example, when the battery is fully charged.

- **Combined Methods**: Combined methods, which account for both voltage and current (as well as other factors such as battery temperature and battery age), can be more accurate than either method alone.

5.4.7 Cell and Battery Voltages

Although it isn't something that the EV driver experiences directly, the voltage of an electric vehicle battery is an important consideration in its design and operation.

Each cell of the battery has a nominal voltage, usually around 3.6 V for a lithium-ion cell—a little more than twice the voltage of AA battery. In order to achieve a reasonable voltage to power a full-size electric vehicle, cells are combined in series, which causes the voltages of the cells to add. For example, combining two 3.6 V cells in series will result in a combined voltage of 7.2 V. Cells are combined in series until the desired voltage is reached, at which point these high voltage series are combined in parallel to increase the available current and power output of the battery pack. As a specific example, the configuration of

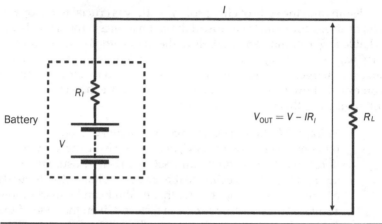

FIGURE 5.3 Internal resistance circuit.

the Tesla Model 3 Long Range battery is reported to be 96s46p, which means 96 cells in series and 46 in parallel.[b]

While each cell, module, and pack has a nominal voltage, also called the open-circuit voltage, the actual operating voltage fluctuates based on whether it is charging (higher voltage) or discharging (lower voltage). This is because each cell has an internal resistance which impedes the flow of current into and out of it (Fig. 5.3).

Following Ohm's law, on discharging we have:

$$V_{OUT} = V - IR_I$$

where V is the open circuit voltage (with no load), I is the current, and R_I is the internal resistance. While the cell provides an open circuit voltage of V, this drops by IR_I over the internal resistance, providing a lower voltage, V_{OUT}, across the terminals.

On charging, the operating voltage increases for the same reason:

$$V_{IN} = V + IR_I$$

To reach a voltage of V across the cell, an additional IR_I must be provided to compensate for the voltage drop across the internal resistance.

5.4.8 Charge and Energy Efficiency

Ideally, a battery would return 100 percent of the energy you put into it. In the real world, however, no battery can do this. It's as if you pour 1 L of water into a bottle and when you go to empty it, less than 1 L comes out. There are two different efficiencies we need to consider: charge efficiency and energy efficiency. Charge efficiency is

---

[b]In actuality, the pack is divided into several different modules with slightly different numbers of cells, but electrically speaking, the overall series-parallel architecture still holds.

the ratio of the amount of Ah you put in to the amount of Ah you get out.

$$\text{charge efficiency} = \frac{Ah_{in}}{Ah_{out}}$$

While other battery chemistries suffer from lower charge efficiencies, the charge efficiency of lithium-ion batteries is generally greater than 99 percent.

Energy efficiency, on the other hand, is the ratio of energy (in Wh) you put into the battery to energy you get out of it.

$$\text{energy efficiency} = \frac{Wh_{in}}{Wh_{out}}$$

Given that batteries have internal resistance, even though greater than 99 percent of the charge you put into a battery comes back out, some of the useful energy conveyed by that charge is lost to the internal resistance, so the total energy efficiency of the battery is lower.

Another reason that batteries can't actually return 100 percent of the energy you put into them is that over time, they slowly discharge themselves. In our water bottle analogy, the bottle has tiny leaks. The rate at which this occurs is called the self-discharge rate. The lower the self-discharge rate, the better. On average, lithium-ion batteries that are not in use lose a few percent of their charge per month. Batteries stored at high temperature have higher self-discharge rates, as do batteries that have been subjected to stress through repeated deep discharging.

5.4.9 Battery Temperature

Battery temperature has an effect on more than just the self-discharge rate. On one hand, lithium-ion batteries operating at low temperature have decreased capacity and power output.[4] On the other, operating a lithium-ion battery at high temperature increases the rate at which its maximum capacity will degrade over time, which decreases battery life.[5]

Therefore, for optimal battery performance, it's important to keep the battery at a moderate temperature. During operation of an electric vehicle, this generally requires a battery cooling system. In a Tesla, this takes the form of a system of coolant lines that snake between the cylindrical cells in the battery modules.

5.4.10 Battery Geometry

Batteries come in many different shapes: just think of a AA vs. a 9-V vs. a watch battery. Lithium-ion batteries generally come in one of three forms: cylindrical cells, prismatic cells, and pouch cells.

All use the same basic construction, with laminated layers of cathode, separator, and anode materials. In a cylindrical cell, this laminated sheet is rolled up like a jelly roll and put inside a hard cylindrical can. In a prismatic cell, the laminated sheet is folded over on itself and put inside a hard polygonal can. In a pouch cell, the folded laminated sheets are contained within a soft pouch. Tesla uses cylindrical cells, because at the time they were developing the original Roadster, these cells had significantly better performance, having undergone a high degree of optimization for prior use in laptops. But other auto manufacturers often use prismatic or pouch cells.

5.5   Electric Vehicle Charging

In order to provide enough energy to keep us driving for more than a few hundred miles, an electric vehicle's battery must be charged. In this section, we'll take a look at electric vehicle charging from a variety of different perspectives.

5.5.1   Charging Level

Electric vehicle charging is divided into several different "levels" of charging depending on the type of voltage involved.

Level 1 Charging

Level 1 charging is a fancy name for plugging your electric vehicle into a regular 120-V household outlet. While it's easy to do, it's not very fast: Level 1 charging adds only a few miles of range for every hour of charging. This means that it would take several days to completely recharge a Model 3 from empty. However, the major benefit of Level 1 charging is that it's available anywhere there's an electrical outlet.

While Level 1 charging is generally considered too slow to be useful for full battery-electric vehicles, it can be sufficient for plug-in hybrids with short electric ranges, e.g., 25 mi for the Toyota Prius Prime. Even if you only ever use Level 1 charging, these short electric ranges can easily be recovered overnight. This is also true for a full battery-electric vehicle as long as you don't drive more miles in a day than you can recover each night.

Level 2 Charging

Level 2 charging refers to charging that occurs at 240 V: the voltage of the washer and dryer circuits in your home. Unlike Level 1 charging where you just plug into any outlet, Level 2 charging requires you to plug into a dedicated charging station (either at home or on the road). While Level 2 charging can reach powers as high as 19.2 kW, it is usually limited to a lower rate (e.g., Tesla's wall connector is limited to 11.5 kW).

This higher power results in much faster charging. At 11.5 kW, the Model 3 can recover 44 mi of range in an hour of charging, and easily charge from 0 to 100 percent overnight.

DC Fast Charging (Level 3)

In both Level 1 and Level 2 charging, the charger feeds AC to an internal charging unit in the car which converts AC to DC and provides DC to the battery. As the voltage, current, and power increase, however, it becomes more efficient to convert AC to DC using an external charging unit which feeds DC directly into the battery, bypassing the internal charging unit. This is known as DC fast charging (sometimes called Level 3 charging). With DC fast charging, the power available is limited only by what the vehicle and charger are rated for, which can range from 50 kW to 250 kW.

Tesla Superchargers are a form of DC fast charging that can provide up to 250 kW of power. Connected to a 250-kW charger, a Model 3 Long Range can charge from empty to 50 percent charge in under 15 min, and from empty to full in under an hour. (See Fig. 5.5 for data from a full charging session.) Tesla has strategically placed their Superchargers so that in most cases, a 30-min charge provides enough range to reach the next Supercharger. This makes it possible to take a Tesla on a road trip anywhere in the contiguous United States.

5.5.2   Charging Connectors

In addition to different charging levels, there are also a variety of charging connectors that are used to plug the charging cable into the car.

SAE J1772

Developed by the Society of Automotive Engineers, the SAE J1772 is the standard connector for electric vehicles in the United States and is used by most electric vehicle manufacturers and charging stations. It can be used for both Level 1 (120-V AC) and Level 2 (240-V AC) charging, with a maximum power transfer of 1.92 kW and 19.2 kW, respectively (Fig. 5.4).

The J1772 has five pins: two for AC power (L1 and N, Line 1 and Neutral), one for ground (PE, Protective Earth), and two for communication between the car and the charger. The proximity pilot (PP) pin tells the car that the charger is connected (preventing the car from driving away), and the control pilot (CP) pin uses a 1 kHz, 12 V square wave to communicate between the charger and car and coordinate the charging process.[c]

The J1772 includes several levels of shock protection. When connected, the pins are isolated on the interior of the connector, ensuring safe charging even in wet conditions. In addition, the design of the connector ensures that the power pins are the first connection made and the last connection broken. When the plug is connected, the ground pin connects first, followed by the power pins, followed by

---

[c]Note that Fig. 5.4 represents the connector at the end of the charging cable. The pins on the car are mirrored horizontally from this, with L1 and PP on the right side.

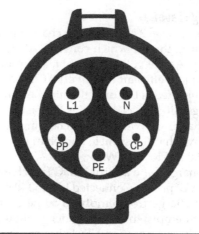

Figure 5.4 SAE J1772 connector.

the communication pins, guaranteeing that power will not flow before the ground and power pins are connected. When the plug is disconnected, the communication pins disconnect first, cutting the flow of power before the power pins are disconnected, preventing arcing, and the ground pin is the last to disconnect, just in case. Furthermore, the proximity pin is connected to a switch on the button that unlocks the connector from the car, cutting the current as soon as the connector is unlocked.

Combined Charging System (CCS)
The Combined Charging System (CCS) is an SAE J1772 connector with two additional pins at the bottom to enable DC fast charging (Fig. 5.5).

IEC 62196 Type 2 (Mennekes)
The IEC 62196 Type 2 connector (commonly known as the Mennekes connector) is the standard for electric vehicle charging in Europe (Fig. 5.6).

The Mennekes connector has seven pins: one for proximity detection, one for control pilot, one for ground, two power pins for charging with single-phase AC power, and two additional power pins for charging with three-phase AC power.

Mennekes CCS
Similar to the J1772 CCS, the Mennekes CCS connector includes two additional pins for DC fast charging (Fig. 5.7).

CHAdeMO
Developed by the CHAdeMO Association (including Nissan, Mitsubishi, and Toyota), the CHAdeMO connector is a dedicated DC fast-charging connector, capable of delivering up to 62.5 kW. It is a popular connector for fast charging in Japan. It includes two power

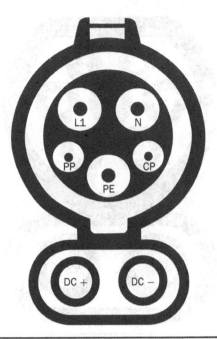

FIGURE 5.5 Combined charging system (CCS) connector.

FIGURE 5.6 Mennekes connector.

pins for DC charging, and eight additional pins for control and communication (Fig. 5.8).

Tesla

Tesla has its own connector which is used to connect to their home chargers and Supercharger network. Although the pins are laid out a little differently, their function is essentially the same as the J1772, with the exception that the Tesla connector can also accept a DC fast

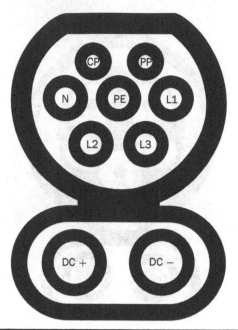

FIGURE 5.7 Mennekes CCS connector.

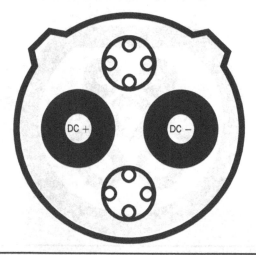

FIGURE 5.8 CHAdeMO connector.

charge. In the United States, Tesla also sells adapters that allow their drivers to connect to J1772, CHAdeMO, and standard wall outlets. In Europe, Teslas are equipped with the Mennekes connector to allow for three-phase charging (Fig. 5.9).

5.5.3 Charging Process

What actually happens when one of the connectors described above is hooked up to an electric vehicle? Earlier in the chapter, we learned

FIGURE 5.9   Tesla connector.

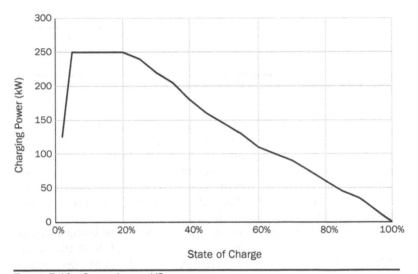

FIGURE 5.10   Supercharger V3 power curve.

that on a microscopic scale, charging a lithium-ion battery involves transporting electrons from the cathode to anode through the circuit, which causes corresponding Li^+ ions to be transported from the cathode to the anode through the electrolyte.

Electrically speaking, this is accomplished by applying a voltage to the battery that is slightly higher than the battery's voltage, which provides the necessary electromotive force to transport electrons from the cathode to the anode.

On a macroscopic scale, here's Fig. 5.10 shows what happens when you plug an electric vehicle into a charging station. This particular curve is based on data from a Model 3 Long Range charging session at one of Tesla's new 250-kW Superchargers, in which the vehicle charged from 2 percent to 100 percent.[6] While the power levels involved are atypical for your average electric vehicle charging session, the general shape of the curve is universal for lithium-ion batteries.

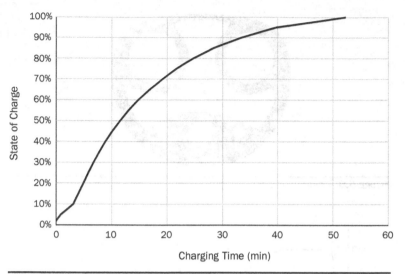

FIGURE 5.11 Supercharger V3 charging time.

At the beginning of the charge session, the car and charging station communicate about what they can each give and receive, and the voltage and current ramp up to a maximum. In the beginning, the battery accepts a large amount of current, recovering energy quickly, but as the charging session continues, the current it accepts begins to decrease. Given constant voltage, this means the power ($P = IV$) also decreases, and with it, the charging rate.

To understand why, remember how lithium-ion batteries work: on charging, lithium ions are intercalated in graphite at the anode. When the battery is at 2 percent charge, most of the space reserved for lithium in the graphite is empty, so it's easy to push lithium ions into it. However, as that space begins to fill up, it becomes more difficult, and the rate at which lithium ions can enter the anode decreases. Given that lithium ions and electrons travel in pairs, and the rate of electron travel is the current, as lithium transport slows, so does the current.[d]

If we plot the same data in a slightly different way, using charging time as our x-axis instead of state of charge, we can clearly see the impact of the decreasing current and power flow. As Fig. 5.11 illustrates, the initial high current from the charger boosts the state of charge to 50 percent quite quickly (in the case of this state-of-the-art Supercharger, it took only 11.5 min). But as the current tapers off, so does the rate at which charge is recovered. In this case, it took more than four times as long to achieve a full charge (52 min) as it did to

---

[d] As an analogy, imagine that the lithium ions entering the anode are people entering a theater. When the theater is empty, it's easy to find a seat, but as the theater fills up, it becomes more difficult, and the rate at which we can get people seated decreases.

achieve a charge of 50 percent. (Of course, 52 min is still quite impressive for a full charge of a car with a range of over 300 mi!)

5.6   Conclusion

In this chapter, we provided an overview of lithium-ion batteries, discussing how they work, what they're made of, and how they can store energy for use in electric vehicles. We also introduced a variety of important characteristics that are used to evaluate batteries in general, and lithium-ion batteries specifically. In addition, we covered electric vehicle charging, including charging levels, charging connectors, and the charging process. Overall, we hope that this chapter has helped you understand what's really going on inside an electric vehicle battery so that in the future, you can see it as more than just a black box that takes power in and puts power back out.

5.7   Homework Problems

5.1 Explain what happens, physically and chemically, when a lithium-ion battery is being charged.

5.2 Explain what happens, physically and chemically, when power is being drawn from a lithium-ion battery.

5.3 Imagine that you are designing an electric vehicle battery pack with approximately 5,000 individual cells (±1%), each with a voltage of 3.6 V. What will the series/parallel configuration (_s_p) of the pack be if you want the final pack voltage to be as close as possible to 300 V? 350 V? 400 V? In each case, exactly how many individual cells will you use, and what exactly will the final pack voltage be? (*Hint*: In real life, you can't have fractional cells in a module, so all of your final values need to be whole numbers—hence the allowance of ±1% in total cells.)

5.4 Early electric vehicles were powered by lead-acid batteries, which currently have an energy density of around 80 Wh/l and 30 Wh/kg. How large and heavy would the 100-kWh battery of the Tesla Model S be if it were made from lead-acid batteries?

5.5 The representation of the Tesla charging connector in Fig. 5.9 has been purposely left unlabeled. Given that the Tesla connector is a proprietary standard, its "pin out" (a diagram showing the arrangement of the pins) isn't widely available. However, given that the function of the five pins is largely the same as the SAE J1772 (aside from allowing DC fast charging), and Tesla provides owners with a simple snap-on converter that allows an SAE J1772 cable to be connected to the car, you can probably make an educated guess. Given your educated guess about the Tesla connector pin out, explain what you think each of the five pins in the charge

FIGURE 5.12   Charging port on a Tesla Model 3.

port of Nick's Model 3 is used for. (*Hint*: Figure 5.12 presents a picture of the car, not the end of the charging cable.)

5.6 As of 2020, the current speed record for the famous Cannonball Run from the Red Ball Garage in New York to the Portofino Hotel in Los Angeles is 26 h and 38 min, at an average (and highly illegal) speed of 106 mph. Using the Tesla Supercharger trip planner (www.tesla.com/trips), determine how long it would take to make the Cannonball Run in the longest range Tesla currently available, assuming that your average speed while on the road matches the record-setting 106 mph. (*Hints*: Your answer will be significantly less than the time quoted by the Tesla trip planner, as that assumes legal speeds. To get your answer, you'll actually need to do some math. In addition, while your real-world range would probably suffer when driving at an average speed of 106 mph, for the purpose of this analysis, you can trust the range estimates used by the trip planner.)

5.7 In ushering in the new era of long-range electric vehicles, Tesla Motors started by introducing a luxury sports car (the Roadster) at a $100,000 price point, then a luxury sedan (the Model S) at a $50,000 price point, and only later released the Model 3 at a $35,000 price point. According to early employees at Tesla, there is a specific reason why the cars had to be released in this order. What do you think this reason might be? (*Hint*: Consider what chapter this homework problem accompanies.)

Notes

1. Trefis Team. "How Battery Costs Impact Tesla's Margins: An Interactive Analysis." *Forbes*, January 13, 2020. https://www.forbes .com/sites/greatspeculations/2020/01/13/how-battery-costs-impact-teslas-margins-an-interactive-analysis/

2. Eure, J. "Scientists Pinpoint the Creeping Nanocrystals Behind Lithium-Ion Battery Degradation." Brookhaven National Laboratory, May 29, 2014. https://www.bnl.gov/newsroom/news.php ?a=24805

3. Lambert, Fred. "Tesla Battery Degradation at Less Than 10% After Over 160,000 Miles, According to Latest Data." *Electrek*, April 14, 2018. https://electrek.co/2018/04/14/tesla-battery-degradation-data/

4. Ji, Y., Y. Zhang, and C.-Y. Wang. "Li-Ion Cell Operation at Low Temperatures." *Journal of The Electrochemical Society*, 160, 4 (2013): A636-A649.

5. Leng, F., C. M. Tan, and M. Pecht. "Effect of Temperature on the Aging Rate of Li Ion Battery Operating Above Room Temperature." *Scientific Reports*, 5, 12967 (2015): 1-12.

6. Holland, Maximilian. "Tesla Model 3 On SuperCharger V3 — Adds 50% Range in Under 12 Minutes!" *CleanTechnica*, June 24, 2019. https://cleantechnica.com/2019/06/24/tesla-model-3-on-super charger-v3-adds-50-range-in-under-12-minutes-charts/

General references for this chapter include:

Pistoia, Gianfranco, and Boryann Liaw (ed.). *Behaviour of Lithium-Ion Batteries in Electric Vehicles: Battery Health, Performance, Safety, and Cost.* Cham, Switzerland: Springer, 2018.

Buchmann, Isidor. *Batteries in a Portable World*, 4th edition. Richmond, BC: Cadex, 2016.

Jiang, Jiuchun, and Caiping Zhang. *Fundamentals and Applications of Lithium-Ion Batteries in Electric Drive Vehicles.* Singapore, John Wiley & Sons, 2015.

Larminie, James and John Lowry. *Electric Vehicle Technology Explained.* Hoboken, NJ: Wiley, 2012.

Image Credits

CHAPTER 6

Controllers

6.1 Introduction

In Chap. 4, we learned about a variety of different electric motors that can be used to turn electricity into rotational motion. In Chap. 5, we learned how lithium-ion batteries can be used to store electricity in a portable form in order to provide power for an electric vehicle motor.

Unfortunately, you can't just wire a motor and a battery together and drive off into the sunset. As we learned in Chap. 5, a battery puts out direct current at a set voltage. With the exception of a brushed-DC motor, which will run at a single speed when connected directly to a fixed voltage, none of the motors in Chap. 4 will work when wired straight to a battery. In order to modulate the speed of a brushed-DC motor, and allow any of the other motor types to function at all, we need to be able to change the characteristics of the electricity coming out of the battery, whether that means converting it to a different voltage, converting it to alternating current, changing the frequency of that alternating current, or all of the above. In this chapter, we'll learn how all of this is done using an electronic tool known as a controller.

6.2 Circuit Elements

We'll get to what exactly a controller is and how it works in a moment, but first, as we did in Chap. 4, let's start by understanding the fundamental components that are required to build one, which are resistors, inductors, capacitors, diodes, and transistors (Fig. 6.1).

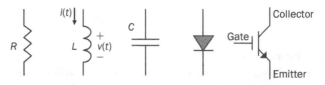

FIGURE 6.1 A resistor, inductor, capacitor, diode, and transistor (from left to right).

125

6.2.1 Resistors

Resistors are used to model elements that convert electrical energy into another form of energy. In this case, we are mostly concerned with the conversion of electrical energy into rotational motion in an electric motor.

Resistors have the symbol R, and their resistance is measured in ohms (Ω). Like all circuit elements, the resistor is characterized by a relationship between the voltage across the element and the current through the element. For the resistor, this relationship is expressed in Ohm's Law:

$$v(t) = i(t)R \qquad (6.1)$$

which states that the voltage, v, as a function of time, is equal to the current, i, as a function of time, times the resistance, R. The voltage has units of joules/coulomb and measures the energy required to move one coulomb of charge through the resistor. The current has units of coulombs/second and measures the rate at which charge is flowing through the resistor.

Since $v(t)$ has units of joules/coulomb and $i(t)$ has units of coulombs/second, the product of the two has units of joules/second, which is power:

$$p(t) = v(t)i(t) \qquad (6.2)$$

If we combine the expression for power in Eq. (6.2) with the relationship between voltage, current, and resistance expressed in Eq. (6.1), we can develop the following equations for power as a function of either voltage or current:

$$p(t) = i^2(t)R \qquad (6.3)$$

$$p(t) = \frac{v^2(t)}{R} \qquad (6.4)$$

If we integrate Eq. (6.2) with respect to time, we get the total energy converted by the resistor (in joules):

$$\mathcal{E} = \int_{t_0}^{t_1} v(t)i(t)dt \qquad (6.5)$$

The average power flow over this period of time is simply the total energy divided by the total time:

$$\mathcal{P} = \frac{1}{t_1 - t_0} \int_{t_0}^{t_1} v(t)i(t)dt \qquad (6.6)$$

6.2.2 Inductors

We've already seen an inductor before, albeit by a different name. The electromagnetic poles in the stator of the motors described in Chap. 4 are inductors. In that chapter, we learned that Biot-Savart's Law can be used to estimate the magnetic field that results from the current

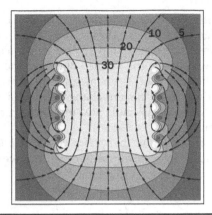

FIGURE 6.2 Magnetic field created by an inductor.

flowing in a coil of wire. Figure 6.2 shows the magnetic field created by a coil made of five loops of wire.

As we saw before, the strength of the magnetic field created is proportional to the current flowing in the loop. This can be expressed by Eq. (6.7):

$$Li(t) = N\Phi(t) \tag{6.7}$$

where L is the inductance of the coil (in henries), $i(t)$ is the current flowing through the coil as a function of time, N is the number of turns in the winding, and $\Phi(t)$ is the resulting magnetic flux as a function of time.

If we change the geometry of coil, for example, by increasing the diameter of the loops, increasing the distance between the loops, or using square loops rather than circular ones, this will modify the inductance, L. Even so, $i(t)$ will remain a linear function of $\Phi(t)$ (and vice versa).

If we take the derivative of Eq. (6.7), we get:

$$L\frac{di}{dt} = N\frac{d\Phi}{dt} \tag{6.8}$$

If you recall from Chap. 4, Faraday's Law states that the right side of this equation is equal to voltage as a function of time:

$$L\frac{di}{dt} = v(t) \tag{6.9}$$

Unlike the fundamental relationship for a resistor, which states that the voltage is proportional to the current, in the case of an inductor, the voltage is proportional to the derivative of the current.

Therefore, to find an expression of the current as a function of time, we simply integrate Eq. (6.9):

$$i(t) = \frac{1}{L} \int_{t_0}^{t} v(\tau)d\tau + i(t_0) \tag{6.10}$$

Equation (6.10) illustrates a fascinating quality of inductors, that is, that the change in the current through an inductor must be continuous: it cannot instantaneously step up or down to a new value. Why is this? Even if we change the voltage abruptly, the current doesn't change immediately. Instead, as an integral function of the voltage, it ramps up over time. If the current changed immediately, this would imply that there was a measurable area under the curve when $dt = 0$, which isn't possible.

From Eq. (6.2), we know that $p(t) = v(t)i(t)$. Given this, and the fact that energy is the integral of power over time, we can say that the energy stored in an inductor's magnetic field is:

$$\mathcal{E}_L = \int_{-\infty}^{t} v(\tau)i(\tau)d\tau$$

$$= \int_{-\infty}^{t} L\frac{di}{d\tau}i(\tau)d\tau$$

$$= \frac{Li^2(t)}{2} \tag{6.11}$$

This further reinforces our inability to instantaneously change the current because it shows that the energy in the magnetic field is proportional to $i^2(t)$. Therefore, if $i(t)$ instantaneously jumped to a new value, the energy would change by some amount, $d\mathcal{E}$, in zero time. But given that $p(t) = d\mathcal{E}/dt$, this means that it would require infinite power to make this change, which isn't possible. As we'll see later, this quality of inductors is one of the essential electromagnetic principles that allows controllers to function.

6.2.3   Capacitors

A simple capacitor with two parallel plates is depicted in Fig. 6.3. Importantly, these two plates are separated by a layer of insulation, so that an electric field can exist across this layer, but charge itself can't flow through it.

The battery depicted in the right side of the figure pulls electrons from the top plates into the positive terminal of the battery. The electrons gain energy as they pass through battery and are deposited on

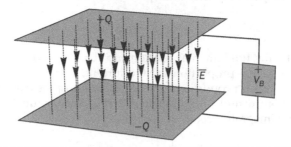

FIGURE 6.3   A parallel plate capacitor.

FIGURE 6.4 A cylindrical capacitor.

the bottom plate of the capacitor. As a result, an electric field is created with the field lines connecting the positive charge on the top plate to the negative charge on the bottom plate.

For a parallel plate capacitor, the charge is proportional to the voltage as given in Eq. (6.12):

$$Q = CV \tag{6.12}$$

in which Q is the charge, V is the voltage, and C is the capacitance (in farads), as defined by the relationship in Eq. (6.13):

$$C = \epsilon_0 \frac{A}{d} \tag{6.13}$$

where A is the area of either plate (in square meters), d is the distance separating them (in meters), and ϵ_0 is the electric constant introduced in Chap. 4. A large capacitance would therefore seem to require a large area. In practice, many capacitors efficiently store this area by rolling the two conducting plates and the insulator that separates them into a cylinder as shown in Fig. 6.4.

While real-world capacitors have a variety of different geometries, the basic relationship $Q = CV$ still holds. The capacitance, C, changes based on the geometry of the device, but the relationship between charge and voltage is still linear.

We can express this relationship as a function of time as given in Eq. (6.14):

$$q(t) = Cv(t) \tag{6.14}$$

If we take the derivative of that, we get:

$$\frac{dq}{dt} = C\frac{dv}{dt} \tag{6.15}$$

Given that current is the change in charge over time, we can say that:

$$i(t) = C\frac{dv}{dt} \tag{6.16}$$

Equation (6.16) gives the relationship between current and voltage for a capacitor. Notice that this is similar to, but nearly the opposite of, the same relationship for an inductor. In the case of an inductor, $v(t)$ was proportional to di/dt, whereas in the case of a capacitor, $i(t)$ is proportional to dv/dt.

Using the same logic that we did in the previous section, we can also say that:

$$v(t) = \int_{t_0}^{t} i(\tau)d\tau + v(t_0) \tag{6.17}$$

In addition, we can find the energy stored in the electric field between the plates as given in Eq. (6.18):

$$\begin{aligned} \mathcal{E}_C &= \int_{-\infty}^{t} i(\tau)v(\tau)d\tau \\ &= \int_{-\infty}^{t} C\frac{dv}{d\tau}v(\tau)d\tau \\ &= \frac{Cv^2(t)}{2} \end{aligned} \tag{6.18}$$

Everything we said in the previous section regarding the inability to change the current instantaneously also applies here, but in this case, it applies to the voltage.

6.2.4 Diodes

A diode is like a one-way valve for current. Roughly speaking, current can flow through it in one direction but not in the opposite direction. More accurately, a diode has the voltage-current relationship shown in Fig. 6.5. As illustrated, the actual voltage-current relationship squiggles through the three regions shown in Fig. 6.5.

The forward region, seen on the right side of the figure, is the normal operating region when the anode is at a higher voltage than the

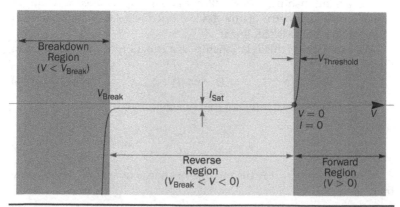

FIGURE 6.5 IV curve for a diode.

cathode. Current is positive, which means it flows forward through the diode. EV-class diodes can carry hundreds of amps in the positive direction. While the voltage drop across the diode is close to zero, a small threshold voltage remains, which means there is at least a small amount of power loss (or perhaps quite a bit, when the current is measured in hundreds of amps, as it is in an EV).

The reverse region, in the center of the figure, is the normal operating region when the anode is at a lower voltage than the cathode. Very little current flows in this case, as long as the diode doesn't break down. EV-class diodes can hold off hundreds of volts in the reverse direction. However, while the current flow is close to zero, a small saturation current does flow in the reverse direction, which means that there is also a small amount of power loss in this case (or perhaps quite a bit, when the voltage is measured in hundreds of volts, as it is in an EV).

The left side of the figure shows the breakdown region. In this case, the diode has failed and become a short circuit in the reverse direction. Diodes can survive a temporary breakdown, but they can also get too hot and fail permanently because in the breakdown region, the current and voltage are both high (in the negative direction). To prevent this from happening, diodes are mounted on a heat sink to sweep away high temperatures.

6.2.5 Transistors

The transistor illustrated in Fig. 6.6 is an insulated gate bipolar transistor (IGBT), which acts as an electronic switch. If the gate-emitter voltage is at or above a certain threshold voltage, current flows from the collector to the emitter.

In the controllers described in the rest of this chapter, a pulse of varying width is applied to the gate. The width of the pulse is controlled to allow just the right amount of energy to flow from the battery into the remainder of the circuit, where it is stored in an inductor or capacitor or both, then smoothly converted to rotational motion by the motor.

EV-class IGBTs wrestle with similar problems to those described in the previous section and illustrated in Fig. 6.5. We want the threshold voltage to be low and the breakdown voltage to be high, and the allowable on-state current needs to be in the hundreds of amps. Generally, IGBTs are arranged in parallel so that they can share the

FIGURE 6.6 An insulated gate bipolar transistor (IGBT).

load of carrying the high on-state current. Finally, the time it takes to switch the transistor on and off needs to be short, because the timing requirements are measured in microseconds.

6.3   Controllers

The term controller is a very broad term of art, covering a wide variety of devices, but the general operation of a controller is illustrated in Fig. 6.7. In general, the feedback system measures, and then actuates (i.e., changes) the state of some "plant," which basically just means anything that has a state that can be changed.

As an analogy, imagine that the plant is an electric oven. The chef selects a temperature for the oven: this corresponds to the command at the left of Fig. 6.7. The sensor, in this case a thermometer, measures the temperature of the oven and compares it to the command. Say the temperature in the oven is lower than desired. The error between the current and set temperatures is sent to the controller, which actuates the oven to heat up, in this case, by turning on the heating element. Then the process is repeated, as new error measurements are sent to the controller, which turns the heating element on and off as necessary to approximate the set temperature as closely as possible.

In the case of this chapter, the plant will be an electric vehicle. While an electric vehicle contains many different control systems, one of the most important relates to its speed. Using the accelerator pedal, the driver commands a certain speed. If the desired speed is higher than the current speed, the controller actuates the motor to speed up.

In a DC motor, this means increasing the voltage at the motor terminals, which increases the current flowing in the windings, and thus, the torque. Given that the battery provides nearly constant voltage, a DC motor controller needs to be able to modulate this voltage up or down in order to increase or decrease the vehicle speed.

In an AC motor, to increase the speed, we need to increase the frequency of the AC waveform that is applied to the motor terminals. Given that the battery only provides DC, an AC motor controller first needs to be able to generate AC waveforms, or at least reasonable approximations thereof. Then, to control the speed of the vehicle, it needs to be able to modulate the frequency of these waveforms.

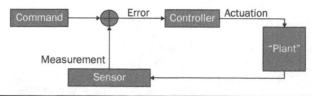

Figure 6.7   General design of a control system.

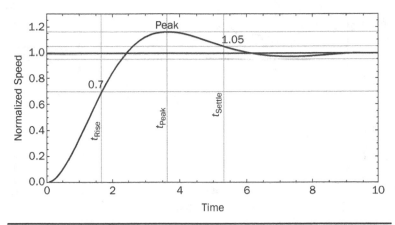

FIGURE 6.8 Response to a step change of magnitude 1.0.

In either case, we want the change in speed to meet certain performance goals, as illustrated in Fig. 6.8. This figure illustrates what happens as a result of a command to increase the speed by 1 mph.

The key performance objectives for this change are as follows:

Rise Time The speed should swiftly converge on the desired value, which in this case is 1 mph faster than the initial speed. In other words, the gap between the initial value (measurement) and the desired value (command) should be closed quickly.

Overshoot However, the more quickly the gap is closed, the more likely we are to overshoot the desired value. In Fig. 6.8, we see an overshoot of approximately 20 percent, to a peak value of 1.2 mph. We also want to minimize this overshoot.

Settling Time After overshooting, we want the speed to decrease (or settle) to the desired value quickly, rather than staying overshot for a long time.

Steady-State Error Finally, once we reach the desired state, we want the continuous, steady-state error to be low. In Fig. 6.8, we see that once the speed is within 5 percent of the desired value, it always stays within 5 percent of the desired value, even though it undershoots the desired value on its way back down from the initial overshoot.

Figure 6.9 illustrates a more realistic scenario in which the variable in question is a noisy one. In the case of speed, this noise might be as a result of an uneven road surface, or uncertainties in measuring the speed of the vehicle. Nevertheless, we still want the four performance objectives described above to hold true.

6.3.1 Step-Down DC Controllers

We are now ready to study a simple DC controller, the step-down DC controller, which can be used to create voltages that are lower than the

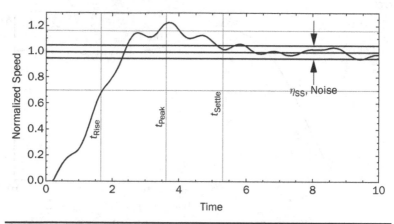

FIGURE 6.9   Noisy response to a step change of magnitude 1.0.

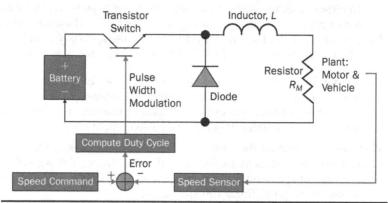

FIGURE 6.10   Block diagram of a step-down DC controller.

battery voltage. This will allow us to vary the voltage applied to a DC motor, which, as we learned in Chap. 4, will allow us to vary its speed.

The controller shown in Fig. 6.10 converts the error between the measured speed and the desired speed into commands for the IGBT transistor switch at the top of the diagram. (As mentioned before, in the high-current environment of an electric vehicle, this will actually be an array of IBGTs working in parallel to share the load.)

If the present speed is too low, the controller increases the fraction of the time that the switch is closed (i.e., on). If the present speed is too high, the controller increases the fraction of the time that the switch is open (i.e., off). In other words, the controller modulates the width of the "on" pulse relative to the "off" pulse. For this reason, we call this process pulse width modulation (PWM).

When the switch is closed (i.e., on), current flows clockwise around the outer loop at the top of Fig. 6.10. This loop is redrawn on the left side of Fig. 6.11. When the switch is closed, the battery provides power to the inductor, L, and the resistor, R_{Mot}, where the latter models our

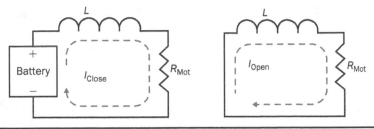

FIGURE 6.11 The two states of a step-down DC controller.

motor. The motor converts some of the power to rotational motion, and the inductor stores some of the power in its magnetic field.

When the switch is open (i.e., off), the battery is no longer connected, and doesn't provide any immediate power to the rest of the circuit, which is now completed by the diode shown in the middle of Fig. 6.10.

However, there is still energy stored in the inductor's magnetic field, which continues to provide power to the motor as shown in the right side of Fig. 6.11. If you recall from Sec. 6.2.2, the inductor current is continuous, i.e., it cannot change instantaneously, so even though the current from the battery has disappeared, current continues to flow from the energy stored in the inductor, powering the rest of the circuit. This continues until the energy stored in the inductor is used up, or the switch is closed and the battery provides new energy.

The voltage that appears across the motor depends on the fraction of time the switch is closed. This fraction is known as the duty cycle, and is calculated by Eq. (6.19):

$$D = \frac{T_C}{T_C + T_O} \qquad (6.19)$$

where D is the duty cycle (a ratio), T_C is the time the switch is closed, and T_O is the time that the switch is open. Their sum, $T = T_C + T_O$, is the total cycle time of the controller. DC controllers typically use $T \approx 50 \times 10^{-6}$ sec. This short time corresponds to a frequency of $f = 1/T$ or around 20 kHz, which is just above the range of human hearing.

The average voltage that appears across the motor can be approximated as:

$$V_{\text{Mot}} \approx D V_{\text{Bat}} \qquad (6.20)$$

This is illustrated in Fig. 6.12, which shows inductor current (left scale) and motor voltage (right scale) for two duty cycles: 80 percent and 20 percent. The waveform on top has an 80 percent duty cycle and converges to an average motor voltage that is 80 percent of V_{Bat}, which is 400 V. The waveform on the bottom has a 20 percent duty cycle and settles at around 20 percent of 400 V. The waveforms in Fig. 6.12 are called sawtooth waves, for obvious reasons. They have one tooth for every T sec, where T is the length of one cycle (approximately 50×10^{-6} s).

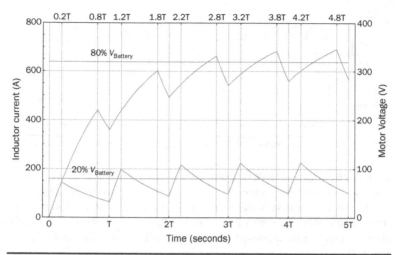

FIGURE 6.12 Motor voltage provided by two different duty cycles.

Take a close look at the first T sec of the waveform for the 80 percent duty cycle at the left side of the figure. From $t = 0$ to $0.8T$, the switch is closed, and the battery is connected. At $t = 0$, the current and voltage are both zero, but they both grow as the battery provides power to the inductor and the motor. At time $0.8T$, the switch is opened, and the battery is disconnected. Now the only source of power is the energy stored in the inductor. Hence, the voltages and currents shown in Fig. 6.12 decay as the energy in the magnetic field is transferred to the motor and converted to rotational power. For an 80 percent duty cycle, this decay is short, lasting only from $0.8T$ to T, then the battery is connected again, and the cycle repeats.

The 20 percent duty cycle is similar, except that the battery is powering the circuit for only the first 20 percent of the cycle, and the inductor powers the circuit for the other 80 percent.

By simply changing the amount of time the switch is open, we can reasonably approximate a wide variety of different voltages, allowing us to run a DC motor at a wide variety of different speeds.

6.3.2 Step-Up/Down DC Controllers

The step-down DC controller described in the previous section allows us to generate voltages that are less than the battery voltage. If we want to generate voltages that are greater than the battery voltage, we can also do that, but it requires a slightly more complicated circuit, one which includes both an inductor and a capacitor.

Figure 6.13 shows a step-up/down controller circuit. As illustrated at the left side of Fig. 6.14, when the switch is closed, energy is transferred from the battery to the inductor. During this period of time, the motor is powered by energy stored in the capacitor.

When the switch is open, as illustrated at the right side of Fig. 6.14, the battery is cut off from the circuit, and the energy stored in the

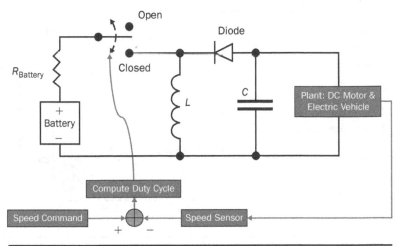

FIGURE 6.13 Block diagram of a step-up/down DC controller.

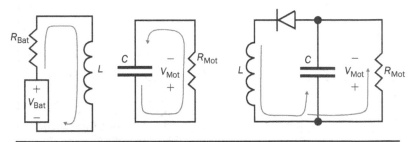

FIGURE 6.14 The two states of a step-up/down DC controller.

inductor provides power to both the motor and the capacitor. Thus charged, the capacitor has energy to power the motor during the next time when the switch is closed.

Figure 6.15 shows the transient, i.e., immediate, response for a step-up/down controller with a duty cycle of 30 percent. In other words, the switch is closed for the first 30 percent of each cycle. During this time, the inductor current grows quickly as the battery charges up the inductor's magnetic field. The capacitor voltage sags during this period because the energy in the capacitor is being used to power the motor. During the remaining 70 percent of each cycle, the switch is open, and the inductor provides energy to both the capacitor and the motor. Therefore, the inductor current sags and the capacitor voltage grows.

Figure 6.16 shows the motor voltage, which is the same as the capacitor voltage, for two different duty cycles: 70 percent and 30 percent. In the 70 percent case, 70 percent of the time is spent building current in the inductor, while the capacitor is providing power to

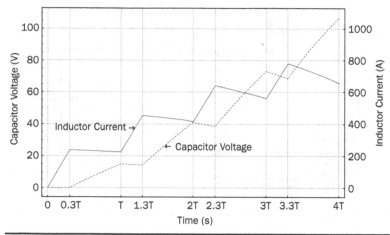

FIGURE 6.15 Transient response of the inductor current and capacitor voltage.

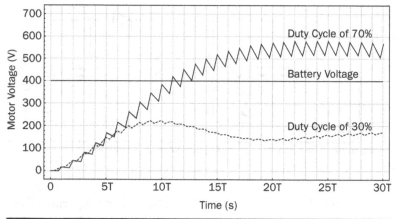

FIGURE 6.16 Steady-state response of the motor voltage for two different duty cycles.

the motor. However, because it's being charged up for so long, the inductor stores much more energy, which boosts the capacitor voltage to higher and higher levels when the switch is finally opened. Figure 6.16 shows 40 cycles, during which the motor voltage reaches steady state at a voltage higher than the battery voltage.

With a 30 percent duty cycle, the inductor stores far less energy, which means the capacitor only ever reaches a fraction of the full battery voltage.

In this way, step-up/down DC controllers are able to create not only voltages that are lower than the battery voltage, but also voltages that are higher than the battery voltage.

6.3.3   AC Controllers

While DC controllers are an essential first step for building our understanding, most electric vehicles in the market today use some kind of AC motor. Therefore, in this section, we'll learn about AC controllers and how they can be used to modulate the speed of AC motors.

The fundamental circuit underlying of an AC controller is illustrated in Fig. 6.17. This kind of controller is called an H bridge because of its distinctive geometry with four switches forming an H around the motor.

Recall that AC motors run off of a sinusoidal voltage function, with the rotor following the peak of the sinusoid around the stator. If we want the vehicle to go faster, we need to increase the frequency of the sinusoid, and if we want it to go slower, we need to decrease the frequency.

To generate a sinusoid, the AC controller first needs to be able to create both positive and negative voltages. In Fig. 6.17, if switches S1 and S4 are closed and S2 and S3 are open, current flows from left to right through the motor, as illustrated by the dashed line which represents current flow. But if switches S2 and S3 are closed, and S1 and S4 are open, current flows from right to left. This allows us to generate voltages of both signs across the motor.

In order to approximate the actual shape of the sinusoid, we use PWM, just as we did in the DC controller. As we learned in Sec. 6.3.1, we use short pulses to approximate low voltage, and long pulses to approximate high voltage. Therefore, when the amplitude of the voltage sinusoid is low, we use short pulses, and when the amplitude is high, we use long pulses. The rectangles in Fig. 6.18 show the varying pulse widths needed to approximate the sinusoid below.

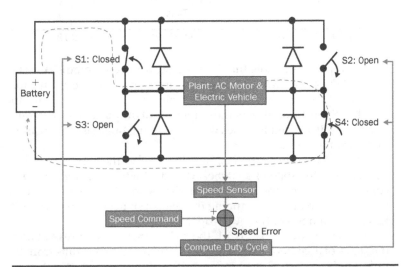

FIGURE 6.17 AC controller based on an H bridge.

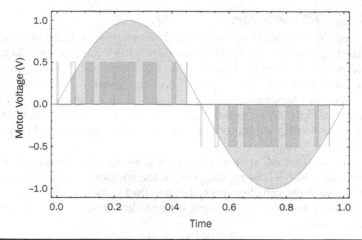

FIGURE 6.18 Pulse width modulation for AC control.

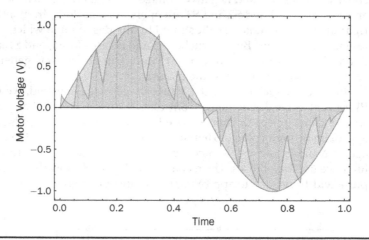

FIGURE 6.19 Circuit response as a result of the pulses in Fig. 6.18.

It's important to note that in this case, we're only modulating the width of the pulses. The H bridge circuit described above has no ability to modulate the amplitude (i.e., voltage) of the pulses, only the length of them. (Modulating the amplitude of the pulses would require a DC controller wired in series with this circuit, but that isn't required in this example.)

The pulses in the first (i.e., positive) half of the cycle are created by changing the state of switches S1 and S4 while leaving S2 and S3 continuously open, and the pulses in the second (i.e., negative) half of the cycle are created by changing the state of switches S2 and S3 while leaving S1 and S4 continuously open.

Just as we saw in Sec. 6.3.1, the pulses create circuit responses in the shape of a sawtooth. The sawtooth waveform that results from the pulses in Fig. 6.18 is shown in Fig. 6.19. As you can see, the sawtooth waveform roughly approximates a sine wave.

By varying the width of the pulses, we can approximate a sinusoid of pretty much any frequency we need, which allows us to change the speed of the motor. Given that we started with a constant DC voltage from the battery, AC controllers are pretty magical devices.

6.4   Conclusion

At the beginning of this chapter, we set up a problem: you can't just connect a motor to a battery and expect it to provide useful work to move a vehicle. In the case of a brushed-DC motor, this would simply cause it to run a single speed, and for all of the other motors we've studied, they wouldn't even function. Controllers solve this tricky problem for us by allowing us to generate a wide variety of voltages, allowing us to run our motor at variable speeds.

The step-down DC controller allows us to create a variety of DC voltages less than the battery voltage, and the step-up/down DC controller allows us to create a variety of voltages that are less than or greater than the battery voltage. Both of these types of controllers can be used to modulate the speed of a DC motor. Finally, AC controllers allow us to transform DC voltage into AC waveforms with variable frequencies, allowing us to modulate the speed of an AC motor.

6.5   Homework Problems

6.1 The Tesla Model 3 Performance has a battery voltage of 350 V and a peak motor output of 340 kW. If we model the motor system as a single resistor, what is the approximate equivalent resistance of the motor system at peak output? What is the current at peak output?

6.2 In your own words, explain why the current through an inductor must be continuous and cannot instantaneously step up or down to a new value. Then, explain why the same is true for voltage in a capacitor.

6.3 Imagine a parallel plate capacitor comprised of two plates the size of 8.5" by 11" sheets of paper, separated by 1 mm. If the electric constant ϵ_0 is equal to 8.854×10^{12} F/m, what is the capacitance of this capacitor? (*Hint*: Be careful with units.) What are two different ways we can increase the capacitance of this capacitor?

6.4 Germanium-based diodes typically have a threshold voltage of around 0.3 V. If 500 amps is flowing through a diode with a threshold voltage of 0.3 V, how much power is being consumed by the diode?

6.5 Before transistors, electric vehicle speed was controlled by connecting a variable amount of resistance in series with the motor (either through a rheostat with variable resistance, or a mechanical switch that selected between one of several different resistances/speeds). Draw a circuit diagram of such a system, and explain how it works. What is the major disadvantage of using such a system? (*Hint*: Assume the motor is a brushed-DC motor.)

6.6 Many luxury vehicles, including Teslas, employ rain-sensing windshield wiper systems that adjust the speed of the wipers as needed based on the amount of water hitting the windshield. Using your knowledge of controllers, explain how such a system works.

6.7 Assuming a step-down DC controller connected to a 350-V battery, what is the duty cycle required to step down to 300 V? 200 V? 100 V? 50 V?

6.8 In your own words, explain how a step-up/down DC controller works.

6.9 Draw a circuit diagram of a control system capable of turning constant DC voltage from a battery into AC voltage with both variable voltage and variable frequency. (*Hint*: You'll need to combine two different types of controllers.)

Notes

General references for this chapter include:

Franklin, Gene, J. David Powell, and Abbas Emami-Naeini. *Feedback Control of Dynamic Systems*, 4th edition. Englewood Cliffs, N.J.: Prentice Hall, 2002.

Johnson David E., Johnny R. Johnson, and John L. Hilburn. *Electric Circuit Analysis*, 3rd edition. New York: John Wiley and Sons, 1999.

Scott, Donald. *An Introduction to Circuit Analysis: A Systems Approach.* New York: McGraw-Hill, 1987.

Kuo, Benjamin. *Automatic Control Systems*, 8th edition. New York: John Wiley & Sons, 2003.

Hayt, William, Jack Kemmerly, and Steve Durbin, *Engineering Circuit Analysis*, New York: McGraw-Hill, 2012.

"Specification Sheet for the FGY120T65SPD F085 650 Volt, 120 Amp, Field Stop Trench Insulated Gate Bipolar Transistor with Soft Fast Recovery Diode." ON Semiconductor, May 2017.

Dodge, Jonathan, and John Hess. "Insulated Gate Bipolar Transistors, Application Note APT0201 Rev. B." Bend, OR: Advanced Power Technology, July 1, 2002.

Image Credits

Figures 6.1 to 6.19: Per Enge

CHAPTER 7

Efficiency and Emissions

So far, we have studied the technology required to build electric vehicles, including motors, controllers, and batteries. But we haven't yet addressed an essential question: are electric vehicles something we actually want to build? In other words, are electric vehicles good for the world? The fact that they can accelerate quickly certainly makes them fun to drive, but how do they compare to other vehicles when it comes to environmental impact, particularly in terms of energy efficiency and greenhouse gas emissions? These are the questions we will address in this chapter.

7.1 Energy Efficiency

In the early days of Tesla Motors, co-founders Martin Eberhard and Marc Tarpenning conducted a well-to-wheels energy analysis of electric vehicles in order to confirm (for themselves and for the public) that they were on the right track in pursuing battery-electric vehicles— as opposed to hybrid, flex-fuel, or fuel-cell vehicles.[1] In this section, we will walk you through a similar well-to-wheels energy analysis, presenting a "back-of-the-envelope" approach to calculating well-to-wheels efficiency of vehicles.[a] Like Tesla, we will focus on the vehicles that currently account for the highest percentage of transportation energy use, i.e. cars.

Before we begin, let's lay out some general assumptions that will remain the same throughout all of our calculations. If you are doing your own calculations, it will be easy to swap out these numbers and use your own.

---

[a]While there are a variety of complex tools for making such calculations, for example, the GREET model (which stands for Greenhouse Gases, Regulated Emissions, and Energy Use in Transportation), we find that a "back-of-the-envelope" style calculation can often be more instructive. This is because all of the assumptions are laid bare and can easily be changed on the fly if you want to use different numbers.

7.1.1 Vehicle Efficiency

For battery-to-wheels efficiency, we will rely on the U.S. Environmental Protection Agency's (EPA) fuel economy ratings for several representative electric vehicles. EPA fuel economy ratings for EVs are given in the form of two numbers: "miles per gallon gasoline equivalent" (mpg-e), which is meant to allow for rough comparisons between the efficiency of EVs and ICEs (although many argue that it's a misleading metric that should be retired), and "kWh per 100 mi," which is a more logical measure to use when we're working with EVs.

Specifically, for a best-case scenario for EVs, we will use the most efficient EV, the 2020 Tesla Model 3 Standard Range, which uses 24 kWh per 100 mi, meaning that it can travel 4.2 mi/kWh. For a more typical scenario, we'll use a number that is representative of an EV somewhere in the middle of the pack and say that a typical EV can travel about 3.2 mi/kWh. This is the efficiency of the Kia Soul Electric as well as several Tesla Model S and Nissan LEAF models.

7.1.2 Charging Efficiency

Before an electric car can go anywhere, its batteries must be charged. Unfortunately, charging is not a lossless process. Studies of real-world electric vehicle charging applications have found that there is generally a loss of around 15 percent from power outlet to car battery due to a variety of inefficiencies in the charging process.[2] This is a loss that will need to be factored into our well-to-wheels EV calculations.

7.1.3 Transmission and Distribution Efficiency

Taking another step closer to the "well," the charger draws its power from the electric grid. Electricity drawn from the grid is subject to transmission and distribution (T & D) losses on its journey from power plant to power outlet, largely as a result of electrical resistance in the wires. T & D losses in the United States are estimated to be around 5 percent.[3] These losses will also need to be factored into our EV calculations.

7.1.4 Generation Efficiency

Before electricity reaches the grid, it is generated through the conversion of a primary energy resource, such as coal, natural gas, or solar. Where relevant, the efficiency of these conversion processes will also factor into our calculations.

As they vary widely, we will consider the assumptions for other fuel and vehicle types on a case-by-case basis in the following sections.

7.2 Petroleum

As we have seen before, 97 percent of transportation energy is currently provided by petroleum. In the case of cars, there are two main petroleum-based fuels we can consider: diesel and gasoline.

7.2.1 Diesel

According to the EPA, the 2019 Chevrolet Cruze is the most fuel-efficient mass-production diesel car currently available. Using one gallon of diesel, it travels an average of 37 mi.[4]

Best Case: Diesel

$$1 \text{ gal} \times \frac{37 \text{ mi}}{1 \text{ gal}} = \textbf{37 mi}$$

Compared to the average car on the road that gets 25 mpg, this seems like a reasonable option.[b]

7.2.2 Diesel Electric

Another option would be to put that same gallon of diesel into a diesel generator and convert it to electricity to power an electric vehicle. The question is: how many miles could this diesel-powered electric car go?

An online diesel generator store is currently advertising a high-efficiency Slow-Turning 5-kW Yanmar Diesel Generator that consumes 0.34 gal/h to generate a continuous output of 4.5 kW, for an efficiency of 35 percent.[c]

Say we have a gallon of diesel: how far can an electric car run on the electricity produced by our generator? From the generator to the battery, the electricity is subject to an electric vehicle charging efficiency of 85 percent. (If we're running the generator at home, right next to the car, we don't have to consider T & D losses from the grid.) The most efficient electric vehicle available today, the 2020 Tesla Model 3 Standard Range, can travel 4.2 mi/kWh. Putting this all together, we find that a diesel-generator-powered Model 3 could travel 48 mi on a gallon of diesel—30 percent farther than the diesel-powered Chevrolet Cruze.

Best Case: Diesel Electric

$$1 \text{ gal} \times \frac{38.3 \text{ kWh}}{1 \text{ gal}} \times 0.35_{\text{Generation}} \times 0.85_{\text{Charging}} \times \frac{4.2 \text{ mi}}{1 \text{ kWh}} = \textbf{48 mi}$$

Why is this hacked-together diesel-electric setup so much more efficient than a finely tuned diesel automobile? One reason is that a stationary diesel generator can be operated at maximum efficiency the whole time it's running, as opposed to a diesel car engine, for which the operating point is constantly changing and often far from optimal. For this reason, many trains and ships have been designed to run as diesel-electric hybrids in which a diesel engine is run at high efficiency

---

[b]Note, however, that a gallon of diesel contains 13 percent more energy than a gallon of gasoline, which means that 37 mpg diesel is actually equivalent to 32 mpg gasoline in terms of pure energy efficiency.
[c]Given that diesel has an energy content of 38.3 kWh/gal, this means that if you run the generator for an hour, you put in 0.34 gal × 38.3 kWh/gal = 13 kWh to get out 4.5 kWh, for a diesel-to-electricity efficiency of 35 percent.

to generate electricity that is then used to drive an electric motor that propels the vehicle.

Another benefit of our diesel-powered electric vehicle over a traditional diesel-powered vehicle is that while the diesel electricity generation still creates emissions, the emissions are all coming from the stationary generator. If we scale this model up to power multiple cars, instead of the emissions coming out of the tailpipes of multiple diesel cars, there is one centralized "tailpipe" for all of the cars, located at the stationary generator. This makes the emissions easier to treat and has the potential to reduce pollution above and beyond the benefits obtained by simply using less fuel.

Of course, that's only the best-case scenario. What about a more typical case? According to the Energy Information Administration (EIA), the efficiency of a typical diesel-powered electricity generation facility was 30.7 percent in 2018.[5] From there, the electricity must travel through the grid, where it is subject to transmission and distribution losses of around 5 percent, followed by a charging efficiency of 85 percent. Therefore, putting average diesel-fired electricity into an average EV with battery-to-wheels efficiency of around 3.2 mi/kWh results in a typical diesel-to-EV efficiency of 30 mpg of diesel.

Typical Case: Diesel Electric

$$1 \text{ gal} \times \frac{38.3 \text{ kWh}}{1 \text{ gal}} \times 0.307 \text{ Generation} \times 0.95 \text{ Grid}$$

$$\times 0.85 \text{ Charging} \times \frac{3.2 \text{ mi}}{1 \text{ kWh}} = \textbf{30 mi}$$

While this isn't as efficient as the Chevrolet Cruze, it's still 20 percent better than the median efficiency of 2019 model year diesel in the United States, which was 25 mpg.

Therefore, if we want to use diesel as a fuel for our cars, it's better, in both the best case and a typical case, to use that diesel to generate electricity to charge an electric car than it is to use it to directly power a car with a diesel engine. While we're not necessarily recommending diesel as a fuel for electric vehicles, if it's what we happen to have available, an electric vehicle is the most efficient way for us to turn it into miles traveled.

7.2.3 Gasoline

Interestingly, what's true for diesel is not true for gasoline. Given that gasoline generators have an efficiency of around 20 percent, and gasoline has an energy density of 37.3 kWh/gal, a best case gasoline-generator-powered Model 3 will only travel 27 mpg.

Best Case: Gasoline Electric

$$1 \text{ gal} \times \frac{37.3 \text{ kWh}}{1 \text{ gal}} \times 0.20 \text{ Generation} \times 0.85 \text{ Charging} \times \frac{4.2 \text{ mi}}{1 \text{ kWh}} = \textbf{27 mi}$$

While this is slightly more than the average fuel economy for gasoline cars (25 mpg), it's nowhere near the fuel economy of the best gasoline-powered car, the Mitsubishi Mirage, with a fuel economy of 39 mpg, and it's less than half the fuel economy of the best hybrid, the Hyundai Ioniq Blue, which gets 58 mpg.[6]

Therefore, if we want to use gasoline to power our vehicles, the best way to do it is through existing ICE vehicles, whether they are purely gasoline-powered, or better yet, hybrid.

7.3 Coal

At the turn of the twentieth century, 40 percent of cars were powered by steam. Of course, steam, like electricity, is only an energy carrier—a fuel source must be burned to generate the steam. Although the energy to boil water could theoretically be generated by any fuel, early steam engines often ran on coal.

7.3.1 Fischer–Tropsch (F-T) Diesel

While steam-powered cars are unlikely to make a comeback anytime soon, there is another method that some people have proposed we could use to turn coal into a transportation fuel again: the Fischer–Tropsch (F-T) process. Invented by German chemists Franz Fischer and Hans Tropsch in 1925, the Fischer–Tropsch process is a chemical method that can be used to convert coal to liquid hydrocarbons, which can be further refined to produce transportation fuels like diesel.

Imagine we have 10 kg of coal that we want to turn into diesel. Stored within 1 kg of coal, there is approximately 6.3 kWh of energy. In a gallon of diesel, there is 38.3 kWh.[7] The F-T process for coal-to-liquids is thought to have a theoretical efficiency of 60 percent and a maximum practical efficiency of around 50 percent.[8] From there, the efficiency of refining the resulting crude oil to usable diesel fuel is around 90 percent.[9] As we have seen before, the most fuel-efficient diesel car has a fuel economy of 37 mpg.[10] Putting this together, a F-T coal-to-diesel car could go 27 mi on 10 kg of coal.

Best Case: Fischer–Tropsch Diesel

$$10 \text{ kg coal} \times \frac{6.3 \text{ kWh}}{1 \text{ kg coal}} \times \frac{1 \text{ gal diesel}}{38.3 \text{ kWh}} \times 0.5_{\text{ Fischer–Tropsch}}$$
$$\times 0.9_{\text{ Refining}} \times \frac{37 \text{ mi}}{1 \text{ gal}} = \mathbf{27 \text{ mi}}$$

7.3.2 Coal-Fired Electric

Another option would be to burn the coal in a high-efficiency power plant to generate electricity to charge an EV. If this were the case, how far could it go?

As a best-case scenario, the Nordjylland Power Station in Denmark runs the most efficient coal-fired generator in the world, with a coal-to-electricity efficiency of 47 percent.[11] From there, the electricity is subject to the usual transmission and distribution efficiency of 95 percent, charging efficiency of 85 percent, and a battery-to-wheels efficiency of 4.2 mi/kWh in the Model 3. Putting this together, we see that an electric car could go 100 mi on 10 kg of coal, almost four times the distance of the best-case F-T diesel car.

Best Case: Coal-Fired Electric

$$10 \text{ kg coal} \times \frac{6.3 \text{ kWh}}{1 \text{ kg coal}} \times 0.47 \text{ Generation} \times 0.95 \text{ Grid}$$

$$\times 0.85 \text{ Charging} \times \frac{4.2 \text{ mi}}{1 \text{ kWh}} = \textbf{100 mi}$$

That's the best-case scenario, but what about a more typical case? According to the EIA, the efficiency of a typical coal plant in the United States was 32 percent in 2015.[12] Putting average coal-fired electricity into an average EV with battery-to-wheels efficiency of around 3.2 mi/kWh results in a typical coal-to-EV efficiency of 52 mi per 10 kg, meaning that our *average* coal-fired EV can still travel almost twice as far as the *best* F-T diesel car, let alone an average one.

Typical Case: Coal-Fired Electric

$$10 \text{ kg coal} \times \frac{6.3 \text{ kWh}}{1 \text{ kg coal}} \times 0.32 \text{ Generation} \times 0.95 \text{ Grid}$$

$$\times 0.85 \text{ Charging} \times \frac{3.2 \text{ mi}}{1 \text{ kWh}} = \textbf{52 mi}$$

Therefore, if we want to use coal as a fuel for our cars, it's better, in both the best case and a typical case, to use that coal to generate electricity to charge an electric car than it is to turn it into liquid fuel to power a diesel car.

7.4 Natural Gas

Although still a fossil fuel, natural gas is often proposed as a cleaner alternative to petroleum and coal for a variety of applications, including transportation.

7.4.1 Compressed Natural Gas (CNG)

One option for using natural gas as a transportation fuel is to compress it and burn it in an internal combustion engine. (While there are currently no consumer CNG cars for sale in the United States, many commercial fleet vehicles run on CNG, so it's still worth comparing the efficiency of this option.) A therm of natural gas is defined as 100,000 BTU (or 29.3 kWh). An optimistic estimate for the efficiency of natural gas compression is around 90 percent.[13] According to the EPA, a previously available CNG car, the 2015 Honda Civic CNG, had a fuel efficiency of 0.92 mi/kWh.[14] Therefore, using a therm of natural gas, a CNG vehicle could travel 24 mi.

Best Case: Compressed Natural Gas

$$1 \text{ therm} \times \frac{29.3 \text{ kWh}}{1 \text{ therm}} \times 0.9 \text{ }_{\text{Compression}} \times \frac{0.92 \text{ mi}}{1 \text{ kWh}} = \textbf{24 mi}$$

7.4.2   Hydrogen Fuel Cell

Alternatively, we could take this therm of natural gas and convert it to hydrogen for use in a fuel-cell vehicle (FCV). The process of converting natural gas to hydrogen is about 75 percent efficient, and compressing hydrogen is about 90 percent efficient.[15] Combined with the 67 mpg-e (2.0 mi/kWh) fuel economy of the 2019 Toyota Mirai FCV, this results in final efficiency of around 40 mi per therm of natural gas, 53 percent better than the compressed natural gas vehicle.[16]

Best Case: Hydrogen Fuel Cell

$$1 \text{ therm} \times \frac{29.3 \text{ kWh}}{1 \text{ therm}} \times 0.75 \text{ }_{\text{Conversion}} \times 0.90 \text{ }_{\text{Compression}} \times \frac{2.0 \text{ mi}}{1 \text{ kWh}} = \textbf{40 mi}$$

7.4.3   Natural Gas Electric

Even better, however, would be to convert natural gas to electricity in a high-efficiency power plant and use this electricity to power an electric vehicle. The most efficient natural gas powered generator in the world, at the Chubu Electric Nishi-Nagoya power plant in Japan, has achieved an efficiency of 63 percent.[17] From there, the electricity is subject to the usual transmission and distribution efficiency of 95 percent, the usual charging efficiency of 85 percent, and a battery-to-wheels efficiency of 4.2 mi/kWh in the Model 3. This results in a best-case efficiency of 62 mi per therm for a natural-gas-powered electric vehicle, 55 percent farther than the fuel-cell vehicle, and more than twice as far as the CNG vehicle.

Best Case: Natural Gas Electric

$$1 \text{ therm} \times \frac{29.3 \text{ kWh}}{1 \text{ therm}} \times 0.63 \text{ }_{\text{Generation}} \times 0.95 \text{ }_{\text{Grid}}$$
$$\times 0.85 \text{ }_{\text{Charging}} \times \frac{4.2 \text{ mi}}{1 \text{ kWh}} = \textbf{62 mi}$$

What if we put electricity from an average natural gas generator into an average electric vehicle? According to the EIA, the efficiency of a typical natural gas power plant in the United States in 2015 was 44 percent.[18] Putting average natural-gas-fired electricity into an average EV with battery-to-wheels efficiency of around 3.2 mi/kWh results in a typical natural gas to EV efficiency of 33 mi per therm, which is 27 percent better than the best-case compressed natural gas vehicle, but not quite as good as the best-case FCV.

Typical Case: Natural Gas Electric

$$1 \text{ therm} \times \frac{29.3 \text{ kWh}}{1 \text{ therm}} \times 0.44_{\text{Generation}} \times 0.95_{\text{Grid}}$$

$$\times 0.85_{\text{Charging}} \times \frac{3.2 \text{ mi}}{1 \text{ kWh}} = 33 \text{ mi}$$

Of course, the small advantage of the best-case natural-gas powered FCV over the typical natural gas powered electric vehicle needs
to be balanced by the fact that electricity generated from natural gas is
widely available everywhere, while hydrogen filling stations are few
and far between. Therefore, it probably makes more sense to devote
our time and effort to bringing the typical natural-gas-powered electric vehicle closer to the best-case natural-gas-powered electric vehicle
rather than developing a completely new hydrogen infrastructure just
to support something that already isn't the best possible way to use
natural gas to power transportation.[d]

7.5   Biomass

Ultimately, our reserves of fossil fuels are limited, and we will eventually need to find an alternative transportation fuel for the future.
One possible option is to use biomass, which can replace fossil fuels
in many applications, including transportation.

7.5.1   Cellulosic Ethanol

One way we can turn biomass into a transportation fuel is to convert
it into the liquid fuel ethanol. A reasonable estimate for ethanol production is around 1 gal of ethanol per 10 kg of biomass.[19] The most
efficient mass-production vehicle that can run on ethanol is currently
the Mercedes CLA250, which gets 20 mpg of ethanol.[20] Putting these
together, a CLA250 could go 20 mi on 10 kg of biomass.

Best Case: Cellulosic Ethanol

$$10 \text{ kg biomass} \times \frac{1 \text{ gal ethanol}}{10 \text{ kg biomass}} \times \frac{20 \text{ mi}}{1 \text{ gal}} = 20 \text{ mi}$$

7.5.2   Biomass Electric

Alternatively, we could take this same 10 kg of biomass and burn it
in a power plant to generate electricity to charge an electric vehicle.
A reasonable estimate for biomass electricity production is 10 kWh
per 10 kg of biomass.[21] Putting this electricity into a Model 3 with an
efficiency of 4.2 mi/kWh, it could go 34 mi per 10 kg of biomass,

---

[d]It doesn't even need to be the best-of-all-possible electric vehicle scenarios to be
better. If you take an average electric car and run it on best-case natural gas electricity, you get 48 mi per therm, and if you take the best electric car and run it on
average natural gas electricity, you get 43 mi per therm, both of which are better
than the 40 mi per therm for the fuel-cell vehicle.

70 percent farther than the best ethanol-powered vehicle. Even an *average* electric vehicle can go 26 mi per 10 kg biomass, 30 percent farther than the best ethanol-powered vehicle. Therefore, if we want to use biomass to power our vehicles, it's better to burn it to generate electricity for an EV than to turn it into biofuel.

Best Case: Biomass Electric

$$10 \text{ kg biomass} \times \frac{10 \text{ kWh}}{10 \text{ kg biomass}} \times 0.95 \text{ Grid} \times 0.85 \text{ Charging}$$
$$\times \frac{4.2 \text{ mi}}{1 \text{ kWh}} = \textbf{34 mi}$$

Typical Case: Biomass Electric

$$10 \text{ kg biomass} \times \frac{10 \text{ kWh}}{10 \text{ kg biomass}} \times 0.95 \text{ Grid} \times 0.85 \text{ Charging}$$
$$\times \frac{3.2 \text{ mi}}{1 \text{ kWh}} = \textbf{26 mi}$$

7.6   Land

While biomass is the immediate resource we are using to produce ethanol and generate electricity, the ultimate resource we are using is the sun shining on a plot of land. Therefore, it's also instructive to look at the amount of land that would be required to support a certain number of miles driven.

7.6.1   Cellulosic Ethanol

If we take 100 sq. ft of farm land and grow miscanthus on it, we can generate about 30 kg of miscanthus in a year.[22] As we saw before, we can then convert this to ethanol at a rate of 1 gal per 10 kg. Putting this together with the 20 mpg of the best ethanol-powered vehicle today, we can drive 60 mi per 100 sq. ft per year.

Best Case: Acres to Ethanol

$$100 \text{ sq. ft} \times \frac{30 \text{ kg}}{100 \text{ sq. ft}} \times \frac{1 \text{ gal}}{10 \text{ kg}} \times \frac{20 \text{ mi}}{1 \text{ gal}} = \textbf{60 mi}$$

7.6.2   Solar Electric

Alternatively, we could take 100 sq. ft of *desert* land and put a solar photovoltaic installation on it. This would generate about 4,125 kWh of electricity per year.[e] Putting this electricity into a Model 3 with an efficiency of 4.2 mi/kWh, we could drive 13,990 mi per 100 sq. ft per

---

[e]Desert land in the United States gets around 7 kWh/m² of sunshine per day averaged across the year. Given that there are 9.29 sq. ft in a square meter, 365 days in a year, and a conservative estimate for solar panel efficiency is 15 percent, we get 4,125 kWh per 100 sq. ft per year.

year. That's more than 200 times as far as we could drive on ethanol! And it's worth noting again that this is *desert* land, while cellulosic ethanol requires farm land. Even if we use only an average electric vehicle, getting 3.2 mi/kWh, we'll still get 10,660 mi per 100 sq. ft per year.

Best Case: Acres to Solar Electricity

$$100 \text{ sq. ft} \times \frac{4{,}125 \text{ kWh}}{100 \text{ sq. ft}} \times 0.95_{\text{Grid}} \times 0.85_{\text{Charging}} \times \frac{4.2 \text{ mi}}{1 \text{ kWh}} = \textbf{13{,}990 mi}$$

Typical Case: Acres to Solar Electricity

$$100 \text{ sq. ft} \times \frac{4{,}125 \text{ kWh}}{100 \text{ sq. ft}} \times 0.95_{\text{Grid}} \times 0.85_{\text{Charging}} \times \frac{3.2 \text{ mi}}{1 \text{ kWh}} = \textbf{10{,}660 mi}$$

7.7   Renewables

Regardless of exactly where and how it is generated, the best case for the environment is to take renewable energy—in the form of hydro, geothermal, wind, or solar power—and use it to power an electric vehicle. As the energy resources themselves are renewable, the only factors we need to account for in terms of efficiency are the transmission and distribution losses and charging efficiency.

7.7.1   Renewable Electric

If we generate renewable energy at a centralized power plant and distribute it to charge electric vehicles, each kWh of renewable energy allows a Model 3 to drive 3.4 mi.

Best Case: Renewable Electric

$$1 \text{ kWh} \times 0.95_{\text{Grid}} \times 0.85_{\text{Charging}} \times \frac{4.2 \text{ mi}}{1 \text{ kWh}} = 3.4 \text{ mi}$$

As for a typical electric vehicle, with a battery-to-wheels efficiency of 3.2 mi/kWh, we'll end up with a power-plant-to-wheels efficiency of 2.6 mi/kWh.

Typical Case: Renewable Electric

$$1 \text{ kWh} \times 0.95_{\text{Grid}} \times 0.85_{\text{Charging}} \times \frac{3.2 \text{ mi}}{1 \text{ kWh}} = 2.6 \text{ mi}$$

7.7.2   Renewable Fuel Cell

While it's clear that in most cases, generating electricity to power an electric vehicle is usually the best way to use energy to generate miles, there's one more thing that's important to consider: how we store the electricity on the vehicle. Some have argued that instead of using a rechargeable battery, it would be better to use electricity to produce hydrogen by breaking water into hydrogen and oxygen, then use that hydrogen to power a fuel cell. According to these proponents of the renewable hydrogen economy, this would allow us to refuel our vehicles at will, using a hydrogen infrastructure similar to our current gasoline infrastructure and circumventing the time required to charge a battery.

While this sounds like it might be a reasonable idea, energetically speaking, it is terribly misguided. This is because electrolysis is only around 70 percent efficient.[23] This, combined with a compression efficiency of 90 percent and a fuel-cell vehicle efficiency of 67 mpg-e (2.0 mi/kWh), results in a total efficiency of 1.3 mi/kWh of renewable energy, which is half that of an electric vehicle powered by renewable energy.

Best Case: Renewable Fuel Cell

$$1 \text{ kWh} \times 0.7 \text{ Electrolysis} \times 0.9 \text{ Compression} \times \frac{2.0 \text{ mi}}{1 \text{ kWh}} = \textbf{1.3 mi}$$

This means that in the theoretical renewables-to-hydrogen scheme, the water, hydrogen, and fuel cell essentially act as an extremely inefficient battery, taking in electricity, and putting out electricity, with huge (and totally unnecessary) losses along the way.[f]

As we saw before, a fuel cell powered by natural gas is one of the rare cases where another type of vehicle can, in some cases, potentially be more efficient at converting energy than an electric vehicle. But just because fuel cells might be an alright way to convert natural gas into vehicle miles, this doesn't mean that fuel cells are a reasonable way to turn renewable energy into vehicle miles. As the calculation above shows, they definitely aren't. If we ultimately want our cars running on renewable energy, battery-electric vehicles are the way to go!

7.8 Greenhouse Gas Emissions

So far, we've seen that electric vehicles are, in most cases, an efficient way to turn primary fuels into miles traveled. But given that transportation accounts for 29 percent of greenhouse emissions in the United States, it's also important to consider how electric vehicles compare in this regard.[24]

Every year, there's a new headline suggesting that electric vehicles are actually an environmental hazard. In particular, it's a common claim that electric vehicles are worse than other vehicles because they are actually responsible for more greenhouse gas emissions per mile than conventional vehicles. In this section, we'll take a look at this

---

[f]And this doesn't even account for transmission and distribution losses, either in the form of electricity losses, if the hydrogen is made onsite, or the energy required to transport the hydrogen, if it's made offsite.

claim, using numbers that are readily available and formulas that are easy-to-understand (unlike many of the sources that make this claim). While the analysis we will do here is specific to the United States, the formulas we use can be adapted to anywhere else in the world as long as you have the corresponding data for your region.

7.8.1   "Tailpipe" Emissions

Some vehicles, like gasoline-powered cars, actually have tailpipe emissions: carbon dioxide is released by the combustion of fuel in the engine and comes out the tailpipe. Electric vehicles, of course, do not. But for a fair comparison, we should trace the energy used to charge an electric vehicle back to its source and account of the emissions from the smokestack of the power plant. We'll refer to these corresponding smokestack emissions as the "tailpipe" emissions of an electric car and compare them to the actual tailpipe emissions of a conventional car.

According to the EIA, the average CO_2 emissions per kWh of electricity generated in the United States was 449 g in 2018.[25] Based on an efficiency of 4.2 mi/kWh, this means that the Tesla Model 3 is responsible for an average of 107 g of CO_2 per mile driven in the United States.

$$\frac{449 \text{ g}}{1 \text{ kWh}} \times \frac{1 \text{ kWh}}{4.2 \text{ mi}} = \textbf{107 g/mi}$$

How does this compare to the average vehicle on the road? Given that combusting gasoline emits 8,890 g of CO_2 per gallon of gasoline, the average U.S. car with a fuel economy of 25 mpg emits 356 g of CO_2 per mile, more than 3 times as much as a Model 3 charged with average U.S. electricity. Therefore, the average Model 3 in the United States is much better than the average gasoline-powered car in the United States.

$$\frac{8,890 \text{ g}}{1 \text{ gal}} \times \frac{1 \text{ gal}}{25 \text{ mi}} = \textbf{356 g/mi}$$

But how does the Model 3 compare to an extremely efficient vehicle? As of 2020, the highest fuel economy for a pure ICE or hybrid vehicle is 58 mpg for the Hyundai Ioniq Blue. Using the same formula as before, we see that the Ioniq Blue emits 153 g of CO_2 per mile, which is 43 percent more than the Model 3. Therefore, the average Model 3 in the United States is much better than the best hybrid in the United States.

$$\frac{8,890 \text{ g}}{1 \text{ gal}} \times \frac{1 \text{ gal}}{58 \text{ mi}} = \textbf{153 g/mi}$$

What would the fuel economy of a ICE vehicle need to be such that it had less emissions than the Model 3? Based on "tailpipe" emissions, the Model 3 running on average U.S. electricity is equivalent to an ICE vehicle that gets 83 mpg of gasoline, which is higher than that of any gasoline-powered or hybrid vehicle ever produced.

$$\frac{8,890 \text{ g}}{1 \text{ gal}} \times \frac{1 \text{ mi}}{107 \text{ g}} = \textbf{83 mpg}$$

Similarly, we could calculate how inefficient an electric vehicle would need to be in order to produce more "tailpipe" emissions than the Ioniq Blue.

$$\frac{449 \text{ g}}{1 \text{ kWh}} \times \frac{1 \text{ kWh}}{\rule{1cm}{0.4pt} \text{ mi}} = \textbf{153 g/mi}$$

Solving for ___ in the equation above, we get 2.93 mi/kWh, or 34 kWh/mi. Some electric vehicles do have lower efficiency than this, e.g., some versions of the Tesla Model X, the Jaguar i-Pace, the Audi e-Tron, and the Porsche Taycan Turbo. But even the least efficient EV on the market today, the Taycan Turbo, is only responsible for emissions of 220 g/mi, which is still far less than the average gasoline-powered car in the United States (at 356 g/mi). In other words, running on average U.S. electricity, even the worst EV on the road is better than the average gasoline-powered car.

Of course, this is just a national average: what happens at the state level? Based on state-by-state emissions data from the EIA, a Model 3 currently has lower "tailpipe" emissions than the Ioniq Blue in 41 U.S. states (plus D.C.) (Fig. 7.1).[8]

In all states, the Model 3 has significantly lower emissions than the average car. The average car has a fuel economy of 25 mpg, compared to the most carbon-intense state, Wyoming, where the Model 3 is responsible for emissions equivalent to a car with a fuel economy of 39 mpg (Fig. 7.2).

Coincidentally, 39 mpg is currently the maximum fuel economy of a non-hybrid, the Mitsubishi Mirage. So this means that: in all states, the Model 3 meets or beats all non-hybrids; in 9 states it's in the hybrid range; and in 41 states (plus D.C.), it's better than the best hybrid.

If we rerun the calculations using an average electric vehicle, we can say that in all states, the average EV beats the average ICE car; in 12 states (plus D.C.) it's in the hybrid range; and in 31 states, it's better than the best hybrid.

7.8.2 Lifecycle Emissions

Of course, "tailpipe" emissions aren't the only emissions associated with a vehicle. There are also emissions related to the energy that it

---

[8]The states where it doesn't are, in order from lowest emissions to highest: Nebraska (where the Model 3 has emissions equivalent to a vehicle with a fuel economy of 54 mpg), Utah (53 mpg), North Dakota (51 mpg), Hawaii (51 mpg), Missouri (46 mpg), Indiana (46 mpg), Kentucky (44 mpg), West Virginia (42 mpg), Wyoming (39 mpg).

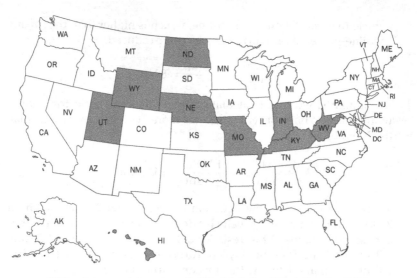

White states = states where the Model 3 wins (41 states plus D.C.)

FIGURE 7.1   Tailpipe emissions of Model 3 vs. Ioniq Blue.

White states = states where the Model 3 wins (all of them)

FIGURE 7.2   Tailpipe emissions of Model 3 vs. average car.

takes to manufacture it. While the energy required to build most components of an electric car and an ICE car is relatively comparable (and therefore, negligible), the battery of an electric car requires a significantly larger amount of energy to produce.

A rough estimate of the emissions produced in the production of 1 kWh of electric vehicle battery (derived by averaging the results from a variety of different analyses) is around 150 kg CO_2 per kWh

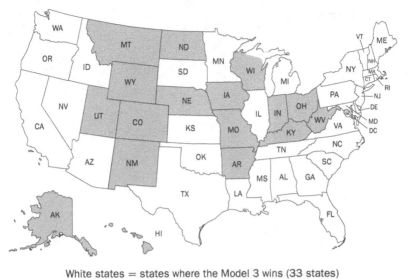

White states = states where the Model 3 wins (33 states)

FIGURE 7.3 Lifecycle emissions of Model 3 vs. Ioniq Blue.

of battery.[26] For a Model 3 with a 54-kWh battery, this amounts to 8,100 kg CO_2. If we divide this by the lower end of the useful battery life advertised by Tesla (300,000 mi), this is equal to 27 g CO_2 per mi.[h] Adding this to the 107 g/mi of tailpipe emissions, we get 134 g CO_2 per mile in the United States, which is still 12 percent less than the emissions of the Ioniq Blue.

On a state-by-state basis, accounting for the battery adds 8 more states (plus D.C.) where the Ioniq Blue has lower emissions than the Model 3, as illustrated in Fig. 7.3.

But still, even in Wyoming, the emissions of the Model 3 are equivalent to a car with a fuel economy of 35 mpg, meaning that in all states, the Model 3 is still significantly better than the average car on the road (Fig. 7.4).

If we rerun the calculations using the Kia Soul Electric, with average EV efficiency of 3.2 mi/kWh and a 64-kWh battery, it is still better than the average ICE in all states, and better than the best hybrid in 20 states.

7.8.3 Future Emissions

Before we get discouraged by the fact that in 17 states (plus D.C.), the Model 3 currently has higher lifecycle emissions than the Ioniq Blue,

---

[h]If this estimate of useful life sounds high, consider that based on a survey of real-world driver data, the batteries in a Tesla were estimated to maintain 90 percent of their capacity for over 180,000 mi. And according to lab simulation by Tesla, a Tesla battery pack maintained 80 percent of its capacity after 500,000 mi worth of cycling.[27]

White states = states where the Model 3 wins (all of them)

FIGURE 7.4   Lifecycle emissions of Model 3 vs. average car.

we should consider what is possible for each option in the future.[i] No matter what we do, given that it is still running on gasoline, the Ioniq Blue will always have tailpipe emissions of around 150 g CO_2 per mile. There's nothing we can do to reduce that, except perhaps some marginal improvements in efficiency in future models (although the maximum fuel economy of hybrids has always hovered at around 60 mpg).

For the Model 3, on the other hand, there's quite a bit we can do to reduce emissions further. On the consumer side, we can charge our electric car using solar energy, which in many places is already less expensive than buying power from the grid. If we don't account for the energy required to manufacture the solar panels, then the 107 g CO_2 per mile from charging the car goes to zero. If we do include lifecycle emissions of around 50 g CO_2 per kWh produced by residential solar panels, our "tailpipe" emissions are still only 12 g CO_2 per mile, which is less than a tenth of the tailpipe emissions of the Ioniq Blue.

Therein lies the real promise of electric vehicles. While ICE vehicles are locked into having the same tailpipe emissions for the life of the vehicle, the "tailpipe" emissions of an electric vehicle will

---

[i]We should also consider that in 33 states, the Model 3 is already a clear winner. And that on average, across the United States, it's also a clear winner, so if we had the chance to replace all cars in the United States with Model 3s or Ioniq Blues, we should choose Model 3s. It's easy to forget that fact, but important not to. It's also important to remember that in every single state, the Model 3 has lower lifecycle emissions than the average car on the road.

become progressively lower over time as the grid becomes less carbon-intensive, and they can also be reduced by around 90 percent at any time simply by charging with renewable energy. In this way, electric vehicles are the only kind of vehicles that have the potential to become progressively more eco-friendly as they age.

On the manufacturing side, we can also use renewable energy to produce the batteries, as Tesla has already begun to do. While some of the emissions associated with producing the batteries will still remain for the foreseeable future (they're not all as a result of electricity use at the factory that can easily be swapped out for solar), we can significantly reduce the lifecycle emissions associated with battery production as well. Of course, even if the manufacturing process doesn't change, if we add together the original estimate of 27 g/mi for battery-associated emissions, and the 12 g/mi of "tailpipe" emissions associated with solar charging (for a total of 37 g/mi), our emissions are still less than a quarter of the emissions of the best hybrid.

7.9   Conclusion

In this chapter, we ran through a series of back-of-the-envelope calculations to see what would happen if we use a variety of different kinds of vehicles to convert a variety of different types of fuel into miles traveled. In doing so, we saw that the most efficient way to turn almost any fuel into miles traveled is to use it to power a battery-electric vehicle.

In addition, we saw that when electric vehicles are powered by actual electricity from the U.S. grid, either as a whole, or in specific states, the average electric vehicle always has less emissions than average car on the road, even after accounting for the energy required to manufacture the battery. Even when pitted against the best hybrid on the road, the Model 3 still wins in 33 states. And while the hybrid will always have the same emissions, electric vehicles will become even cleaner as the power grid continues to decarbonize, and emissions can be reduced even further when cars are charged (and batteries are manufactured) with renewable energy.

Therefore, if our goal is to increase energy efficiency while reducing greenhouse gas emissions, battery-electric vehicles are the way to go.

7.10   Transportation Planning Assignment

To give you a chance to apply what you have learned in this chapter to the real world, we present a long-form assignment in which you will develop transportation plans for several different individuals.

The following people have asked you to be their transportation adviser. Your job is to write a brief transportation plan for each person, in which you tell them how they can best meet their stated goals. As savvy transportation consumers, they won't just take your word for it: you'll need to show them numbers to back up your advice.

If it helps, in designing each plan, you are allowed to use multiple different technologies—the only requirement is that each technology you recommend must be commercially available to the person you're advising, today.

This is an open-ended assignment, meaning that there is no single correct answer to these questions. However, some answers will definitely be more or less correct than others (i.e., if you suggest driving a gasoline-powered Hummer, you'll be wrong).

If you need to assume more than is provided in the scenario (and you will), simply find a source that helps you make a reasonable assumption and cite it.

Scenario 1   Claire lives in Seattle. Like the average driver, she drives approximately 33 mi per day. She likes to lease her cars and has only one concern: keeping her monthly costs low. Her employer recently put in an electric vehicle charging station that is free for her to use. She's intrigued by the possibility of paying $0 for fuel, but doesn't know whether this will actually save her money. Based on monthly cost alone (effective lease price plus current fuel costs), what make and model of car should Claire lease?

Sensitivity Analysis   While Claire currently drives an average of 33 mi per day, it's possible that, in the future, she will drive more or less. At current fuel costs, will your answer change if she drives more or less? For what range of average miles per day (e.g., "10 to 50 miles per day") does your answer hold true, and what should Claire do if she finds herself regularly going outside of this range (at either end)?

Similarly, it's not only likely, but guaranteed, that fuel costs will fluctuate. For what range of fuel costs does your answer hold true, and what should Claire do if fuel costs regularly go outside this range (at either end)? Based on fuel cost trends in Seattle, how likely do you think this is? Finally, while Claire's boss is providing free electricity for now, it's possible that he'll charge for it in the future. At what price per kWh of electricity would your answer change (if any)?

Scenario 2   Per lives in Mountain View and works as a professor at Stanford University. He wants your advice on what car he should buy next. He wants to buy an electric car, but he's not sure this is possible within his constraints:

- His daily commute is 8 mi each way. On the weekends, he drives 30 mi per day.
- Twice a year, he takes a long-distance road trip: (1) to Tahoe, and (2) to Ashland, OR.

- Once a month, he flies out of SFO. He likes to drive himself to the airport and park there.

- He likes to own his cars (vs. leasing), but refuses to pay more than $40,000 to buy one.

- More importantly, he refuses to pay any amount of money for a car that is "ugly" (you can use your own judgment on how to interpret this).

What kind of car should Per buy and why? If he follows your plan, what will the upfront and yearly costs of his transportation be (based on the trips listed above, not including his airfare)? How much energy will he use for ground transportation? How much CO_2 will it emit?

Scenario 3 Tyler lives in Glenwood Springs, Colorado. Working in construction, he needs to drive a truck, which uses quite a bit of fuel. But Tyler is a major fan of the outdoors and still wants to do his part for the environment. Unfortunately, his family makes this difficult, as they are spread around the country: his parents live in Atlanta, GA; his sister in Austin, TX; his grandmother in Bartlesville, OK; and his other grandparents in Snowflake, AZ. He wants to visit each of these places once a year, but he wants to do it in the way that will have the least impact on the environment (note: he wants to physically visit, not just use FaceTime). He's willing to spend a bit more time in transit, and a bit more money, if doing either of these things will help, but he's not quite crazy enough to walk or bike. (He's also aware of the option to use carbon offsets, but he'll feel much better if he can reduce his own emissions as much as possible first.) What mode(s) of transportation should Tyler use to visit his family? And how much CO_2 will he emit in the process?

Notes

1. Eberhard, Martin, and Marc Tarpenning. "The 21st Century Electric Car," October 6, 2006. Tesla Motors.

2. Forward, Evan, Karen Glitman, and David Roberts. "An Assessment of Level 1 and Level 2 Electric Vehicle Charging Efficiency," March 20, 2013. Vermont Energy Investment Corporation. https://www.veic.org/docs/Transportation/20130320-EVT-NRA-Final-Report.pdf

3. EIA. "How Much Electricity Is Lost in Electricity Transmission and Distribution in the United States?" https://www.eia.gov/tools/faqs/faq.php?id=105&t=3

4. EPA. "Fuel Economy Guide." http://www.fueleconomy.gov/feg/findacar.shtml

5. EIA. "Average Operating Heat Rate for Selected Energy Sources." http://www.eia.gov/electricity/annual/html/epa_08_01.html

6. EPA.

7. EIA. "Average Quality of Fossil Fuel Receipts for the Electric Power Industry." https://www.eia.gov/electricity/annual/html/epa_07_03.html

8. de Klerk, Arno. *Fischer–Tropsch Refining*. Weinheim, Germany: Wiley-VCH, 2011.

9. Forman, Grant S. et al. "U.S. Refinery Efficiency: Impacts Analysis and Implications for Fuel Carbon Policy Implementation." *Environmental Science and Technology*, 48 (2014), 7625–7633.

10. EPA.

11. Santoianni, Dawn. "Setting the Benchmark: The World's Most Efficient Coal-Fired Power Plants." https://www.worldcoal.org/setting-benchmark-worlds-most-efficient-coal-fired-power-plants

12. EIA. "Average Operating Heat Rate for Selected Energy Sources." http://www.eia.gov/electricity/annual/html/epa_08_01.html

13. Greenblatt, Jeffery D. "Opportunities for Efficiency Improvements in the U.S. Natural Gas Transmission, Storage and Distribution System." https://www.energy.gov/sites/prod/files/2015/05/f22/QER%20Analysis%20-%20Opportunities%20for%20Efficiency%20Improvements%20in%20the%20U.S.%20Natural%20Gas%20Transmission%20Storage%20and%20Distribution%20System.pdf

14. EPA.

15. Kalamaras, Christos M., and Angelos M. Efstathiou. "Hydrogen Production Technologies: Current State and Future Developments." *Conference Papers in Energy*, 2013. https://www.hindawi.com/journals/cpis/2013/690627/ and Gardiner, Monterey. "Energy Requirements for Hydrogen Gas Compression and Liquefaction as Related to Vehicle Storage Needs." *DOE Hydrogen and Fuel Cells Program Record*, 2009. https://www.hydrogen.energy.gov/pdfs/9013_energy_requirements_for_hydrogen_gas_compression.pdf

16. EPA.

17. Power Engineering. "GE-Powered Plant Awarded World Record Efficiency by Guinness." *Power Engineering*, March 27, 2018. https://www.power-eng.com/2018/03/27/ge-powered-plant-awarded-world-record-efficiency-by-guinness/

18. EIA. "Average Operating Heat Rate for Selected Energy Sources." http://www.eia.gov/electricity/annual/html/epa_08_01.html

19. Nalley, Lanier, and Darren Hudson. "The Potential Viability of Biomass Ethanol as a Renewable Fuel Source: A Discussion." https://www.researchgate.net/publication/23748558_The_Potential_Viability_of_Biomass_Ethanol_as_a_Renewable_Fuel

20. EPA.

21. McKendry, Peter. "Energy Production from Biomass (Part 1): Overview of Biomass." *Bioresource Technology*, 83 (2002), 37–46.

22. McKendry.

23. Stolten, Detlef. *Hydrogen Science and Engineering: Materials, Processes, Systems and Technology.* Hoboken, NJ: John Wiley & Sons, 2016.

24. EPA. (2020). "Sources of Greenhouse Gas Emissions." https://www.epa.gov/ghgemissions/sources-greenhouse-gas-emissions

25. EIA. (2019). "State Electricity Profiles." https://www.eia.gov/electricity/state/

26. Hall, Dale, and Nic Lutsey. "Effects of Battery Manufacturing on Electric Vehicle Life-Cycle Greenhouse Gas Emissions." The International Council on Clean Transportation, February 2018. https://theicct.org/sites/default/files/publications/EV-life-cycle-GHG_ICCT-Briefing_09022018_vF.pdf

27. Lambert, Fred. "Tesla Battery Degradation at Less Than 10% After Over 160,000 Miles, According to Latest Data." *Electrek,* April 14, 2018. https://electrek.co/2018/04/14/tesla-battery-degradation-data/

Image Credits

Figures 7.1 to 7.4: ⊜①◎ Nick Enge, using MapChart.net

CHAPTER 8

Incentives and Barriers

8.1   Introduction

In Chap. 7, we saw that electric vehicles are an effective way to reduce energy consumption and emissions. For this reason, many governments around the world have sought ways to increase their adoption. In this chapter, we will discuss a variety of different policies that have been used at the country, state, and local level to increase the proportion of electric vehicles on the road. In addition, we will discuss several important barriers to EV adoption and show how these barriers are being overcome.

8.2   Incentives

While there are many bottom-up reasons that drivers decide to purchase an electric vehicle (e.g., interest in the technology, concern for the environment, a desire to pay less for automotive fuel and maintenance), in this section, we will focus on policy mechanisms that can be applied from the top down to further encourage EV adoption.

8.2.1   Zero-Emission Vehicle Mandates

Historically, one of the most effective policy mechanisms that governments have used to increase the market share of EVs is to simply mandate that automakers produce and sell them. In the past, even though there was demand for EVs, few automakers were interested in making them in any meaningful quantity, so governments stepped in to give them a push.

In the United States, the Clean Air Act gives California the authority to set emissions standards for vehicles, including the requirement that a certain percentage of vehicles sold must be zero-emission vehicles.[a] Once set, other states can choose to join California's standards. Currently, thirteen states—Colorado, Connecticut, Delaware, Maine, Maryland, Massachusetts, New Jersey, New York, Oregon, Pennsylvania, Rhode Island, Vermont, and Washington—which represent more than a quarter of the U.S. vehicle market, have joined California's Zero-Emission Vehicle (ZEV) mandate.

Under the mandate, automakers are required to produce a number of ZEVs each year, based on the total number of cars they sell in the state (i.e., if your total sales are higher, your sales of ZEVs must be as well). Each vehicle sold receives credits based on its all-electric driving range: the higher the range, the more credits it receives. For example, in 2019, plug-in hybrids received between 0.4 and 1.3 credits per vehicle, based on their all-electric driving range, and all-electric vehicles received between 1 and 4 credits based on their range.

Requirements under the mandate are set as a percentage of credits compared to total vehicle sales, ranging from 7 percent in 2019 to 22 percent in 2025. This means that if you sell 100,000 cars in California in 2019, you need to sell enough ZEVs to reach 7,000 credits. Given that each vehicle can be worth more than one credit, the requirement of 22 percent credits in 2025 will actually translate to less than 8 percent in terms of the number of vehicles sold (as ranges increase, the number of credits per vehicle will increase, and the percentage required in terms of number of vehicles will decrease). As an additional requirement, as of 2019, a maximum of 43 percent of an automaker's credits can come from plug-in hybrids. The remaining 57 percent must come from battery-electric or fuel-cell vehicles.

If an automaker produces more than the required number of credits in a given year, they can bank them for future use, or they can trade or sell them to other automakers that aren't producing the required number of credits. During the first quarter of 2020, Tesla sold a record $354 million of ZEV credits to other automakers.[1] Due to the fact that they exceeded the relatively low requirements set in past years, as of October 2018, automakers had banked enough credits to comply with the mandate through 2022 with no new EV sales.[2]

According to an analysis of the effectiveness of various EV policy mechanisms, signing on to the ZEV mandate is the single most effective thing a state can do to increase the market share of EVs in their state.[3]

In late 2019, the U.S. EPA and the National Highway Traffic Safety Administration (NHTSA) moved to revoke California's waiver under the Clean Air Act to set both ZEV and Greenhouse Gas standards,

---

[a]While this might initially seem unfair, giving one state the authority to set emissions standards avoids the issue of having a patchwork regulatory landscape across the country, in which automakers would need to meet different requirements in every state. California was chosen to be the one state because it has the highest emissions standards.

placing the ZEV program in jeopardy.[4] In November 2019, California and 22 other states filed suit to challenge the U.S. EPA's action to revoke the waiver.[5] This confrontation between federal and state environmental and transportation officials represents a significant escalation over U.S. environmental policy with important consequences for the future of electric vehicle sales. While unresolved as of this writing, experts expect that the case may eventually be heard by the U.S. Supreme Court.

Outside the U.S., many other governments are considering similar ZEV mandates, with some countries proposing an aggressive target of 100 percent of new vehicle sales being electric in the coming decades. (More on this in the next chapter.)

Rather than mandating that a percentage of all vehicles are zero-emission vehicles, some states have mandated that a percentage of fleet vehicles (government fleet, corporate fleet, or both) are zero-emission vehicles. While this is also a reasonable idea, so far, it has yet to make a significant impact on the total market share for EVs in those states.[6]

8.2.2 Fleet Emissions Requirements

Since 2009, national requirements for vehicle efficiency in the United States have been jointly overseen by NHTSA and the EPA. The standard, known as Corporate Average Fuel Economy (CAFE), traditionally applied to gasoline-powered vehicles and provided incentives for automotive manufacturers to decrease the fuel consumption of their vehicles.

However, the standard also includes a number of provisions which create incentives for automakers to manufacture electric vehicles, such as excluding upstream emissions from electricity generation in emissions calculations and offering multipliers which treat the production of one electric vehicle as multiple vehicles, artificially increasing an automaker's measured performance under the CAFE standard.

In addition, in 2018, California passed SB-1014, enacting the Clean Miles Standard for ride-hailing. This policy requires Transportation Network Companies (TNCs) like Uber, Lyft, and other ride-hailing providers to meet as-yet-unspecified maximum CO_2 limits, calculated per mile of passenger travel.

Unlike other policies which regulate vehicle technology or fuel costs, compliance with the Clean Miles Standard will be calculated by measuring both vehicle efficiency (measured in grams of carbon per mile of vehicle travel) and transportation system efficiency (measured by estimating the ratio of vehicle miles traveled and passenger miles traveled). Under the standard, transportation system efficiency is approximated using average ride-hailing occupancy numbers reported by the TNCs.

As of this writing, the Clean Miles Standard rule-making process is underway. The California Air Resources Board and Public Utilities

Commission are responsible for finalizing rules in 2021. Ride-hailing represents a small but rapidly growing fraction of miles driven in many cities, and other states have expressed interest in replicating California's policy, so the impact of policies like this is expected to grow over time.

8.2.3   Rebates and Tax Incentives

In addition to mandating that automakers produce electric vehicles, governments can also help make them more affordable. Rebates, tax credits, and tax exemptions are three different methods that governments have used to reduce the upfront cost of EVs.

With a rebate program, when someone buys an electric vehicle and reports this purchase to the government, the government sends a check to reimburse them for a portion of the purchase (in some cases, the check may be sent to the seller in order to reduce the purchase price from the start). While the EV policy landscape is changing rapidly, many U.S. states have offered rebates on the order of $2,000 for the purchase of a new electric vehicle.

A tax credit is similar to a rebate, except that it is paid out in the form of reducing the taxes you owe in the future. For example, with the U.S. federal electric vehicle tax credit, if you buy an EV in 2019, you'll receive a tax credit of $7,500, which will reduce your federal tax bill when you file your 2019 taxes in 2020.[b]

While similar to rebates in some ways, tax credits have a few caveats: (1) the delivery of the money is delayed, being paid out to you when you file your taxes in the following year, and (2) it can only reduce your tax bill to $0, so you need a tax bill of at least the amount of the tax credit to get the whole benefit. For example, if you only owe $5,000 in taxes, you only get $5,000 worth of the potential benefit, not the full $7,500. Both of these caveats make it more difficult for lower-income families to take advantage of tax credits.

Tax credits can also be offered by state governments, but given that the tax bill from state governments is usually smaller than the tax bill from the federal government, tax credits from state governments are even more limited in size than tax credits from the federal government.

In any case, rebates and tax credits have proven quite effective. According to the previously cited analysis of policy effectiveness, EVs have significantly more market share in states with rebates or tax credits than they do in states that don't.[7]

Another way for governments to reduce the upfront cost of an electric vehicle is to waive the vehicle sales tax on the purchase, which reduces the price of the vehicle by around 5.75 percent on average in

---

[b]Unfortunately for purchasers of EVs made by Tesla and GM, the tax credit is phased out for vehicles from a given automaker once they have sold 200,000 EVs.

the United States (e.g., around $2,200 on the $37,990 base Model 3). In places where the vehicle sales tax is higher, i.e., Norway, which usually has a 25 percent tax on vehicles, but waives this for EVs, the incentive to purchase an EV is even higher.

In addition to helping reduce the cost of electric vehicles, governments can also use rebates and tax incentives to support the installation of charging stations (by homeowners, businesses, or both). In states that have these types of policies, there are significantly more charging stations per capita, which has translated into significantly greater market share for EVs. According to a regression analysis, about half of the increased market share in these states can be attributed to the support they provide for charging infrastructure.[8]

8.2.4 Reduced Vehicle Fees

In addition to reducing the upfront cost of buying an EV, governments can reduce the ongoing costs of owning an EV by reducing or eliminating vehicle fees such as registration and emissions testing fees. While this is certainly one of many tools in the EV policy toolbox, given that the fees aren't very large in the first place (10s or 100s of dollars, rather than 1,000s), reducing them hasn't proven particularly effective at incentivizing EVs: states with reduced vehicle fees for EVs have no greater EV market share than those that don't.[9]

Interestingly, many states have gone the other way and increased registration fees on EVs. This is intended to make up for the money that is lost when the EV driver stops buying gasoline and the state loses gasoline tax revenue. Fortunately, this doesn't seem to be acting as a significant disincentive to buy EVs, as the EV market share in states that have done this is no less than the market share in states that haven't.[10] However, as the proportion of electric vehicles on the road increases, it will be interesting to see what governments do to make up for the increasing loss in fuel tax revenue (for example, California's current revenue from state gasoline tax is over $9 billion.)

8.2.5 HOV Lane Access

Another policy that has been used to incentivize EV purchases is giving electric vehicles the right to travel in HOV (high-occupancy vehicle) lanes, regardless of how many people are in the vehicle (i.e., a solo driver of an EV can join the other vehicles that have the otherwise required two or more people in them). In places where traffic is particularly bad, this can be a strong incentive to switch to an electric vehicle with HOV access.

According to the policy analysis cited above, this has proven to be a powerful incentive for electrification, as states that give EVs HOV lane access have significantly greater EV market share than states that don't.[11]

8.2.6 Support for Charging

When Nick and his wife Melissa bought a Model 3 and wanted to install a home charger, they had to pay several thousand dollars for the installers to run conduit under the eaves in front of their house in order to get power to the garage from the main electrical panel on the other side of the house, then hire someone to paint the conduit to appease their homeowner's association (and their own aesthetic sensibilities).

Much of this time and effort could've been saved if the house was designed to be EV-ready, with an available 240-V circuit pre-wired in the garage. For this reason, some governments have begun to adopt EV-ready wiring codes that require single-family homes, multi-family residences, businesses, and/or parking lots to be pre-wired to accommodate future EV chargers. In addition, some particularly ambitious governments are beginning to require that EV chargers be installed in a certain proportion of parking spaces from the beginning.

Given that 80 percent of charging is done at home, it's often difficult for drivers who live in multi-unit dwellings (e.g., apartments) to consider getting an EV unless the owner of their building decides they want to add a charger. But now, in California, owners of multi-unit dwellings are not allowed to prevent their tenants from adding an EV charger (at the tenant's cost), either in their own designated parking spot, or, if they don't have one, in the common parking area. As the share of EVs on the road increases, the incentive for building owners to install chargers as an amenity to attract potential tenants will increase, and we can imagine that at some point, building owners may even be required to add them.

8.3 Barriers

While there are many different incentives to EV adoption, both natural and policy-driven, there are also several important barriers that must be overcome in order for electric vehicles to achieve greater market share. In this section, we'll discuss the two most important ones, as well as some of the ways that these barriers are being overcome.

8.3.1 Range Anxiety

Range anxiety, which is the fear that an electric vehicle will not have enough charge to get you to your destination, has always been believed to be one of the most significant barriers to electric vehicle adoption. Recently, however, it has been suggested that this fear may be overblown.[12] Before long-range EVs were readily available to purchase, a 2016 study found that even an EV with around 75 mi of range could meet the needs of average U.S. drivers on 87 percent of their travel days, even without charging at all during the day.[13] Even so, the possibility that there might be even one day that your car might not be able to get you where you need to go can be stressful. But it's important to note that this figure from 2016 can now be regarded as conservative in several ways.

First, the newest generation of EVs regularly have over 200 mi of range, which means that they can easily satisfy the demands of many additional travel days. According to the analysis in the 2016 study, an EV with over 200 mi of range would now be able to meet the needs of more than 98 percent of travel days across the United States.

Second, it overlooks the fact that the majority of people (58 percent in 2017) have access to more than one vehicle in their household.[14] As long as every person in the household doesn't independently need to travel more than 200 mi on the same day, the person who does can take a different car (a hybrid, plug-in hybrid, or a conventional vehicle), while the person whose needs that day can be met by the EV drives it.

Finally, in today's transportation landscape, many drivers do have opportunities to charge during the day, given that there are currently over 75,000 (and growing) non-residential EV charging points available across the United States, including many that are at or near the workplace.[15] And as fast charging stations like Tesla Superchargers become more ubiquitous, charging during the day will become even easier, in the rare cases where it's actually necessary.[c]

8.3.2   Cost

Today, EVs are still, on average, more expensive to buy than their ICE counterparts. But this is expected to change very soon. In 2017, Bloomberg New Energy Finance predicted that EVs would reach purchase price parity with comparable ICEs in 2026. In 2018, they revised that prediction to 2024. And in 2019, they revised it once again to 2022, as a result of battery prices dropping even faster than expected.[16]

Of course, it all depends on how much battery an EV is equipped with. Another 2019 analysis suggested that by 2025, an EV with 150 mi of range will be cheaper than a comparable ICE, while an EV with 200 mi of range will be the same price, and an EV with 250 mi of range will still be slightly more expensive. But in the end, even this analysis suggested that the 250-mi-range versions of all three types of vehicles studied—cars, crossovers, and SUVs—will be cheaper than comparable ICEs before the end of the decade.[17] As one article put it, this means that "choosing an electric car over its combustion-engine equivalent will soon be just a matter of taste, not a matter of cost."[18]

Of course, the yearly costs of driving an electric car are already much lower than driving an ICE car. Figures 8.1 and 8.2 show the possible range of per-mile fuel costs for an average ICE car, the most

---

[c] As a side note, it's time to put to bed the old myth that plugging in your electric car to charge at night is "inconvenient," as many people have claimed in the past. Every person we've talked to who has actually owned an EV themselves greatly prefers to spend 5 seconds unplugging their car when they leave and 5 seconds plugging it in again when they return home than to have to remember to stop at the gas station to fill up their ICE car.

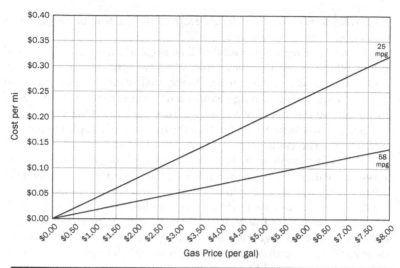

FIGURE 8.1 Cost per mile to fuel a gasoline car.

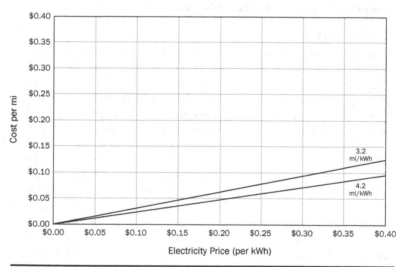

FIGURE 8.2 Cost per mile to fuel an electric car.

efficient ICE car, an average electric car, and the most efficient electric car. The scale of the x-axis has been designed to show all of the possible prices around the world, from $0.02 per gal for gasoline in Venezuela to nearly $8 per gal in Hong Kong, and $0.02 per kWh in Egypt to nearly $0.40 per kWh in Denmark.

While these figures are useful for calculating the cost per mile for an ICE and an EV if you know the price of gas and electricity in a given location, how do these costs compare to each other? Figure 8.3 shows

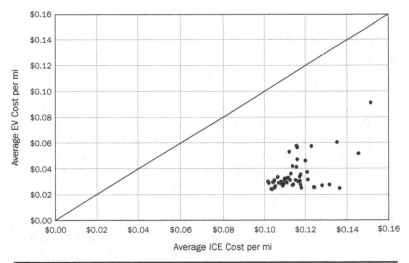

FIGURE 8.3 Fuel cost for EV vs. ICE in the United States.

the fuel cost for an average ICE car and an average EV in the fifty U.S. states (plus DC) in 2018. The diagonal line represents the point at which the fuel cost of an EV and ICE car is the same. As you can see, in every state (each represented by a point), the cost per mile of driving an average EV is significantly cheaper than driving an average ICE car.[19]

In 2018, across the United States, the average fuel cost for an EV was just 30 percent of the average fuel cost for an ICE car, for a savings of 70 percent. In addition, given that EVs have fewer parts and fluids that need replacing, their maintenance costs are significantly lower as well. Estimates for savings on yearly EV maintenance costs compared to ICE maintenance costs have ranged from around one-third to two-thirds.[20] The fact that electric vehicles are so much cheaper in terms of yearly operating costs is one of the many reasons that consumers have already been adopting them even though the purchase price is still a little higher, on average.

8.4 Conclusion

In this chapter, we looked at a variety of different policy mechanisms that have been successfully used to increase electric vehicle adoption, including ZEV mandates, financial incentives, giving EVs special privileges, and providing support for charging.

In addition, we discussed the two major barriers to electric vehicle adoption, namely range anxiety and upfront cost, and described some of the ways in which they are being overcome, specifically, through increased range and decreased battery cost.

8.5   Homework Problems

8.1 Imagine that you are an official at the U.S. DOT or U.S. EPA. What are some of the costs and benefits of a policy to accelerate EV adoption?

8.2 Imagine that you are a lawmaker/regulator at the federal level. What do you think would be the single most impactful policy (or policies) you could enact to accelerate electric vehicle adoption in the United States, and why? Bonus points for coming up with a policy (or policies) that haven't already been discussed. What about if you were a lawmaker/regulator at the state level? What about at the city level?

8.3 In this chapter, we focused on policy mechanisms specifically designed to accelerate the adoption of battery-electric vehicles. Skeptics of electric vehicle policy (as well as other types of technology policy) frequently criticize the government for picking a specific technology to support, when, in their view, the market should be allowed to decide which technology wins, even if they support the overall goal of making passenger vehicles more efficient and lower-emitting. With this in mind, what policies might you enact, at any of the three levels explored above, to achieve this same goal without picking a specific technology winner?

8.4 Consider the following formula (introduced by researcher Lee Schipper), which breaks down the various factors that go into determining the CO_2 emissions associated with transportation:

$$CO_2 = \text{passenger-miles} \times \frac{\text{vehicle-miles}}{\text{passenger-miles}}$$

$$\times \frac{\text{energy}}{\text{vehicle-mile}} \times \frac{CO_2}{\text{energy}} \qquad (8.1)$$

In order to reduce CO_2 emissions from transportation, we can reduce any of the four terms in the formula. In other words, we can (a) decrease passenger-miles traveled, (b) decrease the ratio of vehicle-miles to passenger-miles (which makes more sense to think about as increasing the number of passengers per vehicle), (c) decrease energy consumption per vehicle-mile, or (d) decrease CO_2 emissions per unit of energy. For each of these four possible intervention points, brainstorm several different policies that could be enacted to reduce CO_2 emissions from transportation.

Notes

1. Beresford, Colin. "Other Automakers Paid Tesla a Record $354 Million Last Quarter." *Car and Driver*, May 2, 2020. https://www

.caranddriver.com/news/a32346670/other-automakers-paid-tesla-record-354-million/

2. Union of Concerned Scientists. "What Is ZEV?" https://www.ucsusa.org/resources/what-zev

3. Cattaneo, Lia. "Plug-In Electric Vehicle Policy: Evaluating the Effectiveness of State Policies for Increasing Deployment." Center for American Progress, 2018. https://www.americanprogress.org/issues/green/reports/2018/06/07/451722/plug-electric-vehicle-policy/

4. U.S. EPA. "One National Program Rule on Federal Preemption of State Fuel Economy Standards." https://nepis.epa.gov/Exe/ZyPDF.cgi?Dockey=P100XI4W.pdf

5. Shepardson, David. "California, Other U.S. States Sue to Block EPA from Revoking State Emissions Authority." *Reuters*, November 15, 2019. https://www.reuters.com/article/us-autos-emissions-california/california-other-u-s-states-sue-to-block-epa-from-revoking-state-emissions-authority-idUSKBN1XP25Q

6. Cattaneo.

7. Cattaneo.

8. Cattaneo.

9. Cattaneo.

10. Cattaneo.

11. Cattaneo.

12. Mooney, Chris. "Range Anxiety is Scaring People Away from Electric Cars—But the Fear May Be Overblown." *Washington Post*, August 15, 2016. https://www.washingtonpost.com/news/energy-environment/wp/2016/08/15/range-anxiety-scares-people-away-from-electric-cars-why-the-fear-could-be-overblown/

13. Needell, Z., J. McNerney, M. Chang, et al. "Potential for Widespread Electrification of Personal Vehicle Travel in the United States." *Nature Energy* 1, 16112 (2016).

14. U.S. Federal Highway Administration. "National Household Travel Survey 2017 – Households – Vehicles Available." https://nhts.ornl.gov/households

15. Wagner, I. "Number of Public Electric Vehicle Charging Stations and Charging Outlets in the U.S. as of March 3, 2020." https://www.statista.com/statistics/416750/number-of-electric-vehicle-charging-stations-outlets-united-states/

16. Bullard, Nathaniel. "Electric Car Price Tag Shrinks Along with Battery Cost." *Bloomberg*, April 12, 2019. https://www.bloomberg.com/opinion/articles/2019-04-12/electric-vehicle-battery-shrinks-and-so-does-the-total-cost

17. Lutsey, Nic, and Michael Nicholas. "Update on Electric Vehicle Costs in the United States through 2030." The International Council on Clean Transportation, April 2, 2019. https://theicct.org/sites/default/files/publications/EV_cost_2020_2030_20190401.pdf

18. Bullard.

19. 2018 electricity price data from: EIA. "State Electricity Profiles." https://www.eia.gov/electricity/state/ and 2018 gasoline price data from: Ballotpedia. "Gasoline Costs by State, 2018." https://ballotpedia.org/Gasoline_costs_by_state,_2018

20. Logtenberg, Ryan, James Pawley, and Barry Saxifrage. "Comparing Fuel and Maintenance Costs of Electric and Gas Powered Vehicles in Canada." 2°Institute, September 2018. https://www.2degreesinstitute.org/reports/comparing_fuel_and_maintenance_costs_of_electric_and_gas_powered_vehicles_in_canada.pdf

The Sierra Club and Plug-In America's joint publication *AchiEVe: Model State & Local Policies to Accelerate Electric Vehicle Adoption*, of which Version 3.0 was published in 2019, also served as a useful reference for this chapter.

Image Credits

Figures 8.1 to 8.3: Nick Enge

CHAPTER 9

The Future of Electric Vehicles

9.1 Introduction

As of this writing, more than 10 countries have announced that they intend to ban petroleum-based vehicles in the coming decades, as they imagine a future which is 100 percent electric. Leading the pack, Norway has set a goal for all new cars sold to be zero-emission starting in 2025, and they're making good progress toward that goal: in 2019, over 50 percent of new vehicles sold in Norway were electric.[1] Other countries seem to be settling on a timeline of 10 to 20 years for complete electrification of new vehicles, with many fossil-fuel-powered car bans proposed for 2030 and 2040.

In parallel, automakers around the world are investing heavily in electric vehicles, with many new electric models planned for the coming years. In addition to investing in EVs, several automakers have publicly committed to completely electrifying their product lineups.

Volvo has announced that all new models sold in the future will include some level of electrification, meaning that all cars they sell with be either electric or hybrid.[2] General Motors has gone even further by announcing that they eventually intend to go all-electric by ditching the internal combustion engine entirely.[3] In late 2018, Volkswagen made a similar commitment, adding that "the year 2026 will be the last product start on a combustion engine platform."[4]

Given that electric vehicles are here to stay, let's explore some of the exciting developments that the future might hold for them.

9.2 Charging

While most EV owners are content to plug in their car every night where they park[a] and recover the range they need for the next day while they sleep (or plug in during the day to recover range while

---

[a]According to the U.S. Department of Energy, in the United States, 80 percent of charging occurs at home.[5]

they work), there are several developments on the horizon that could make charging even easier.

9.2.1 Wireless Charging

In recent years, wireless charging has become ubiquitous in the consumer electronics space. Now, instead of plugging in your phone at night, all you need to do is set it down on a charging pad. In the future, the same could also be true for your car. In fact, wireless charging already exists for some electric car models through the Plugless Power wireless charging system. By installing a wireless charging pad in your garage and a wireless charging adapter to the underside of your compatible electric car, you can simply drive it over the pad and your car will begin charging. The Gen 2 system is compatible with the Tesla Model S and BMW i3, and provides 7.2 kW of power at 88 percent efficiency compared to wired charging.[6]

The current problem with wireless car charging is that there isn't yet a universal standard, like the Qi charging standard for consumer electronics, although several groups are trying to establish one. Without such a standard, potential charging pad manufacturers don't have an incentive to design a pad that they have no guarantee car manufacturers will support, and car manufacturers don't have an incentive to design a wireless power receiver into their cars when they have no guarantee there will be a compatible transmitter. Establishing such a standard would make wireless car charging much more feasible.

Once such a standard is in place, one can imagine that cars could be wirelessly charged not only at home (or at work), but anywhere there are currently wired chargers. Rather than having to plug in your car, you could simply park it in a wireless charging spot, run your errands, and be on your way with increased range. Recently, Norway announced its intention to install wireless car chargers for electric taxis in taxi fare lines, where the taxis wait for their turn to pick up customers.[7] While this plan has so far been stymied by the challenges described above (they've yet to find manufacturers willing to partner with them), it's certainly an intriguing idea for the future.

Taking this idea one step further, another futuristic possibility is to charge a specially designed electric vehicle by driving it along a specially designed charging lane on the highway. While the technology is expensive, proponents say that it would be significantly cheaper per mile to install a charging lane than it would be to install a new mass transit system. In addition, they argue that if cars can be charged while they are driving, their batteries can be smaller (requiring only the range required to get to the next charging lane, not to complete the whole trip), and therefore, less expensive. While they will face similar, if not greater, challenges than static wireless chargers, wireless charging lanes are currently being tested in several different countries, including Sweden.[8]

9.2.2 Quick Charging

While some are trying to make charging more convenient by reducing the number of steps required, others are trying to make it more convenient by reducing the time it takes to charge. For example, in 2019, Tesla announced its V3 Superchargers that can deliver a peak charging power of 250 kW. At 250 kW, a Model 3 can charge at a rate of up to 1,000 mi of range per hour, or 75 mi in 5 min.[9] Once its international network of Superchargers is retrofitted to deliver this rate of charging, fueling up your Tesla will become even more similar to filling up your tank at the gas station, and range anxiety will become a thing of the past.[b]

9.3 Battery

While the success of electric vehicles now seems all but guaranteed, the speed with which the transition to them occurs will largely depend on technological and economic developments in the battery space, as the battery of an electric vehicle largely determines its two most important characteristics: its range and its price compared to similar ICE vehicles.

9.3.1 Increased Range

In the 2015 model year, the longest EPA-reported range for an electric vehicle was 270 mi for the Tesla Model S with 90-kWh battery pack. The second longest range that year (outside of other Model S options) was 93 mi for the Kia Soul Electric.

In the 2020 model year, the longest EPA-reported range for an electric vehicle is 391 mi for the Tesla Model S Long Range Plus, which is 45 percent more than the range of the Model S in 2015, and more than four times the range of the Kia Soul Electric in 2015. The second longest range in 2020 (outside of other Tesla models) is 259 mi for the Chevy Bolt. The 2020 Kia Soul Electric now has a range of 243 mi, and several others (the Jaguar I-Pace and Hyundai Kona Electric) have ranges over 200 mi.

Future models are expected to have even longer ranges: the tri-motor Tesla Cybertruck has an advertised range of over 500 mi, and the new Tesla Roadster has an advertised range of 620 mi. With a range of 620 mi, you could drive for nearly 10 h at highway speeds, after which point the driver would probably have an even greater need to recharge than the car!

Increased range is a function of several different things: (1) increased battery density (which allows for more energy to be

---

[b]Previously, several companies, including Tesla, were trying to achieve this goal through battery swapping, in which your entire battery was swapped out in a few minutes, but at this point, interest appears to have shifted firmly, and likely permanently, toward high-power charging.

stored in a vehicle of the same weight and size), (2) decreased cost (which allows for more energy to be stored in a vehicle of the same price), and (3) the willingness to make vehicles with larger batteries (as Tesla has been, and is still, a leader in).

9.3.2 Increased Density

While precise numbers about the energy density of the individual cells in production EVs is closely guarded, we can make a rough estimate of the energy density of the entire pack based on publicly available figures. In 2008, the original Tesla Roadster had a 53-kWh pack that weighed 450 kg, for a pack-level energy density of 106 Wh/kg. In 2017, the Tesla Model 3 Long Range had a 75-kWh pack that weighed 480 kg, for a pack-level energy density of 156 Wh/kg, a 50 percent improvement in under a decade. As battery pack density continues to increase, this will allow for the same range with less weight (and therefore, greater efficiency), or greater range at the same weight and efficiency.

9.3.3 Decreased Cost

According to a recent analysis by Forbes, the estimated cost of the battery in a Tesla has decreased from $230 per kWh in 2016 to $127 per kWh in 2019. Outside of Tesla, Forbes estimates that EV battery costs have fallen from $288 per kWh in 2016 to around $158 per kWh in 2019.[10] As battery costs continue to fall, this will allow for lower-price electric vehicles with the same range, or higher-range electric vehicles for the same price.

9.4 Innovative Vehicle Designs

Electric cars made by the large auto manufacturers are still essentially retrofits of existing models. The basic design of the car hasn't changed, they've just swapped out the ICE transmission for an electric one, putting the new components in place of the old ones. While this is certainly a good first step toward electrification, it misses out on potential innovations that would be possible if, instead of modifying an existing design, they designed a new EV from the ground up.

Tesla's first car, the original Roadster, was a modification of the Lotus Elise, with the ICE components replaced by EV ones. But when Tesla began designing their own cars from the ground up, they were able to be more innovative. For example, instead of placing the motor(s) under the hood, and filling available empty space with batteries, Tesla placed the motor(s) in line with the wheels and used a large flat battery pack as the floor of the vehicle. This means that there's space for a large crumple zone (and a small front trunk, or "frunk") under the hood and that the center of gravity of the vehicle is extremely low, meaning that it's nearly impossible to roll over. In addition, given that the drivetrain is almost entirely in line with the floor, the design of the vehicle above the floor can be pretty much anything they want it to be, as long as it's reasonably aerodynamic. The body of the vehicle can be a sedan (the S and the 3), an SUV (the X), a crossover (the Y), a sports car (the Roadster), or a pickup (the Cybertruck), all riding on essentially the same chassis.

Currently, the designs of the S, 3, X, Y, and Roadster are relatively conservative, mimicking existing high-end vehicles, which makes sense given that the company is trying to prove that its designs can compete with the popular models of the major manufacturers. The Cybertruck, on the other hand, is radically different from any other vehicle that is currently on the market. While one can certainly argue about whether this new design is a good one (there are strong proponents on both sides of the debate), one has to applaud Tesla for taking the risk to break the mold and show that other, radically different designs are possible when you redesign a car from the ground up.

9.5   Toward Complete Electrification

In Chap. 1, we noted that throughout the twentieth century, the residential, commercial, and industrial sectors progressively moved away from the direct burning of fossil fuels toward meeting their energy needs through the use of electricity, a shift which the National Academy of Engineering identified as the "greatest engineering achievement of the 20th century." With the continued growth in popularity of electric vehicles, this trend is likely to continue as the transportation sector joins the other three in electrifying.

As noted, this shift will be a positive one, for several reasons. First, electricity serves as a universal converter of energy, so, as we saw in Chap. 7, we can use a wide variety of energy sources to meet our transportation energy needs, instead of relying almost entirely on petroleum. And second, electricity is a "sticky" form of energy, allowing us to move toward energy resilience (a more realistic goal than energy independence).

As we progress further into the twenty-first century, we can expect that an increasing proportion of this diverse and resilient energy mix will be provided by renewable sources. Some people, including Nick's former academic advisor, Mark Jacobson at Stanford, even believe that we may be able to meet our energy needs with 100 percent renewable energy in the not-so-distant future.[11]

In fact, for individual families, this futuristic dream is available today. The home of Nick and his wife Melissa in Texas is completely electric (like many homes in Texas, it wasn't built with a natural gas line), and the primary vehicle that they drive over 90 percent of their miles in is an all-electric Model 3. To power their home and vehicle, they recently installed solar on their roof, which, after federal tax incentives and local rebates, is providing more than 100 percent of the energy needs of their home and vehicle at a price lower than electricity from the grid or gasoline from the gas station. (And this is in Texas, where both electricity and gasoline are significantly cheaper than the national average!)

9.6   Conclusion

The future of electric vehicles is bright. With both governments and car manufacturers committing to an electric future, we can expect the proportion of electric vehicles on our roadways to continue to increase. At the same time, we can expect them to continue to improve, in terms of cost, range, and convenience of charging. As manufacturers become more comfortable with electric cars, it's possible that we'll see more innovative designs that take advantage of the fact that EVs don't necessarily need to look like existing vehicles. And finally, we can look forward to an increasingly electrified world, where our energy needs are met by a diverse and resilient mix of energy resources, including an increasing proportion of renewables.

9.7   Homework Problems

In 1940, the great automotive pioneer Henry Ford predicted, "A combination airplane and motor car is coming. You may smile. But it will come." Nine years later, the Taylor Aerocar completed a test flight, and in 1956, it was granted certification by the Civil Aeronautics Association (the predecessor of the FAA). Despite these early successes, only six prototype Aerocars were ever built. Today, more than 60 years and countless magazine cover stories later, there isn't a single flying car on the market, although some people still hold out hope that one is right around the corner.

On the other hand, a little over forty years ago, in 1977, Ken Olsen, founder of Digital Equipment Corporation predicted, "There is no reason anyone would want a computer in their home." Today, most of us probably own quite a few computers, and our lives wouldn't be the same without them.

Throughout history, technological predictions, both optimistic and pessimistic, have been proven wrong, sometimes dramatically so. Nevertheless, humans still like to predict things, and sometimes, our predictions are actually proven right.

In the following questions, you will be asked to make a variety of predictions about the future of transportation. While we won't know whether your predictions are right until they come to pass (or don't), the degree to which you use evidence and logic to support your predictions, as opposed to making fanciful guesses, will determine how they are judged in the present.

9.1 What will the transportation landscape look like in 2050? What kinds of vehicles will we use to get around? What proportion of passenger-miles will be traveled in each kind of vehicle? What proportion of the power for each kind of vehicle will be provided by electricity? What about in 2100?

9.2 Consider an electric car in 2050. What will its range be? How will it charge, and how quickly? What features will distinguish it from the average car on the road today? What about in 2100?

9.3 In terms of their popularity in the media, autonomous vehicles are the new flying cars. But unlike flying cars, there actually are autonomous vehicles driving on public roads today, whether they're Waymo taxis, or Teslas on Autopilot. In 2050, what proportion of cars on the road will be (a) fully autonomous, without requiring human supervision, (b) partially autonomous, under human supervision, or (c) primarily driven by humans, maybe with the help of a few driver assistance features like lane departure alerts and early collision warnings? What about in 2100?

9.4 What is something unexpected that you think will happen in the future of transportation? Something that most people would think sounds totally insane today, but that future historians will look back and wonder, "why didn't they see that coming?"

Notes

1. Jones, Harvey. "What's Put the Spark in Norway's Electric Car Revolution?" *The Guardian*, July 2, 2018. https://www.theguardian.com/money/2018/jul/02/norway-electric-cars-subsidies-fossil-fuel

2. Eisenstein, Paul A. "Volvo Is First Automaker to Offer Electric or Hybrid Only." *NBC News*, July 5, 2017. https://www.nbcnews.com/business/autos/volvo-going-all-electric-first-automaker-ditch-combustion-engine-n779791

3. Eisenstein, Paul A. "GM Is Going All Electric, Will Ditch Gas- and Diesel-Powered Cars." *NBC News* October 2, 2017. https://www.nbcnews.com/business/autos/gm-going-all-electric-will-ditch-gas-diesel-powered-cars-n806806

4. Reuters. "Volkswagen Says Last Generation of Combustion Engines to Be Launched in 2026." *NBC News*, December 5, 2018. https://www.nbcnews.com/business/autos/volkswagen-says-last-generation-combustion-engines-be-launched-2026-n943991

5. Office of Energy Efficiency and Renewable Energy. "Charging at Home." https://www.energy.gov/eere/electricvehicles/charging-home

6. Plugless Power. "Plugless Q&A." https://www.pluglesspower.com/plugless-questions/

7. Statt, Nick. "Norway Will Install the World's First Wireless Electric Car Charging Stations for Oslo Taxis." *The Verge*, March 21, 2019.

https://www.theverge.com/2019/3/21/18276541/norway-oslo-wireless-charging-electric-taxis-car-zero-emissions-induction

8. Boffey, Daniel. "World's First Electrified Road for Charging Vehicles Opens in Sweden." *The Guardian*, April 12, 2018. https://www.theguardian.com/environment/2018/apr/12/worlds-first-electrified-road-for-charging-vehicles-opens-in-sweden

9. The Tesla Team. "Introducing V3 Supercharging." *Tesla Blog*, March 6, 2019. https://www.tesla.com/blog/introducing-v3-supercharging

10. Trefis Team. "How Battery Costs Impact Tesla's Margins: An Interactive Analysis." *Forbes*, January 13, 2020. https://www.forbes.com/sites/greatspeculations/2020/01/13/how-battery-costs-impact-teslas-margins-an-interactive-analysis/

11. Jacobson, Mark Z. *100% Clean, Renewable Energy and Storage for Everything*. Cambridge: Cambridge University Press, 2021.

APPENDIX A

Project: Building a Model Electric Car

Nearly 200 years ago, Hungarian priest Ányos Jedlik built the first DC motor complete with stator, rotor, and commutator and demonstrated its potential by building a model electric car.

Your mission, should you choose to accept it, is to retrace Jedlik's footsteps and build your own model electric car. **The only concrete requirement for this project is that your car needs to be propelled by a motor that you build from the ground up with your own hands.**

FIGURE A.1 A successful model car from our class at Stanford.

Having guided many students through this project over the years, there are several pieces of advice we can give you:

- You will have the greatest chance of success if you work in a team. In our Stanford classes, students worked in teams of three or four. The most successful teams were those which included students with a variety of different skill sets, including design, fabrication, and electrical expertise.

- The Internet is your friend. A quick Google search will reveal many different tutorials on how to build an electric motor from scratch. Given the wealth of knowledge available to you, there's no reason to reinvent the wheel. Shop around for a while, considering a variety of different designs, then choose the one that most appeals to you.

- Although some students have started with more ambitious plans, all of the functional model cars that have been presented to us at the end of a 10-week course have used brushed DC motors. This is because every other practical motor design requires designing a system to modulate the electrical waveform as well as designing the physical motor. While not entirely out of the realm of possibility, this has so far proven too daunting a task for an introductory 10-week course for sophomores.

- Note that your ultimate goal is to make a car that moves, at any speed. Trust us when we say that this is enough of a challenge as it is. Although things like aerodynamics matter a lot for a full-scale car, your priority in this project should be the drivetrain.

- On that note, you'll probably need to employ some kind of gear system (albeit with only a single gear), and fiddle with it until you find a reasonable balance between torque and angular velocity. Many of our students have succeeded in building a motor that turns when it has no load but struggled to generate enough torque to move their car forward. Even if your car is as slow as a snail, as long as it moves forward, you've successfully completed your task.

- Aside from generating enough torque to move the car, your greatest challenge will be designing a reliable commutator, so start working on this problem early. Students in the past have struggled mightily with finding a design that ensures the brushes are reliably in contact with the split ring without creating too much friction. On the split ring side of the equation, many students have sung the praises of copper tape.

- While some of our students have machined their own custom parts, others have taken the easier route by building their chassis and drivetrain from existing building sets like LEGO and k'nex. Both approaches are totally fine. As we mentioned at the beginning, the only concrete requirement for this project is that your car be propelled by a motor that you build from the ground up with your own hands. The rest of the car, including the axle and gears, can certainly be designed using existing parts.

- As you learned in Chap. 4, the torque generated by your motor is directly proportional to the number of turns in the windings, so you're probably going to want quite a few of them. A helpful hint that many of our students have passed on is that the winding process can be assisted by an electric drill, which creates the necessary rotary motion much more efficiently than you can yourself. While you are required to build the motor from the ground up with your own hands, those hands are certainly allowed to use power tools.

Image Credit

Figure A.1: Nick Enge

Index

Note: Page numbers followed by f, t, and n indicate figures, tables, and footnotes, respectively.